Compressed Air
Operations
Manual

Compressed Air Operations Manual

An Illustrated Guide to Selection, Installation, Applications, and Maintenance

Brian S. Elliott

McGraw-Hill

New York   Chicago   San Francisco   Lisbon   London   Madrid
Mexico City   Milan   New Delhi   San Juan   Seoul
Singapore   Sydney   Toronto

The McGraw·Hill Companies

CIP Data is on file with the Library of Congress.

Copyright © 2006 by The McGraw-Hill Companies, Inc. All rights reserved. Printed in the United States of America. Except as permitted under the United States Copyright Act of 1976, no part of this publication may be reproduced or distributed in any form or by any means, or stored in a data base or retrieval system, without the prior written permission of the publisher.

1 2 3 4 5 6 7 8 9 0   DOC/DOC   0 1 3 2 1 0 9 8 7 6

ISBN 0-07-147526-5

The sponsoring editor for this book was Kenneth P. McCombs, the editing supervisor was David E. Fogarty, and the production supervisor was Richard C. Ruzycka. It was set in Century Schoolbook by International Typesetting and Composition. The art director for the cover was Brian Boucher.

Printed and bound by RR Donnelley.

This book was printed on acid-free paper.

McGraw-Hill books are available at special quantity discounts to use as premiums and sales promotions, or for use in corporate training programs. For more information, please write to the Director of Special Sales, McGraw-Hill Professional, Two Penn Plaza, New York, NY 10121-2298. Or contact your local bookstore.

Contents

List of Figures and Tables   xi
Preface   xxi
Acknowledgments   xxiii

Chapter 1. The History of Compressed Air   1

  Questions   8

Chapter 2. Compressed Air Applications   9

  Blow-Off Guns   10
  Tire Inflators   12
  Portable Air Tanks   14
  Fire Extinguishers   15
  Air Cylinders   16
  Pressure Amplifiers   25
  Percussion Tools   27
  Air Motors   29
  Rotating Tools   33
  Turbine Motors   37
  Paint Sprayers   38
  Siphon Guns   40
  Sandblasters   42
  Air Nailers   45
  Air Staplers   46
  Caulk Guns   48
  Grease Guns   48
  Pop Rivet Puller   49
  Dual Diaphragm Pumps   50
  Vacuum Generators   51
  Air Bearings   52
  Spray Misters   53
  Vibrators   54

Automobile Lifts   55
Air-Over-Hydraulic Jacks   56
Air Bag Jacks   59
Pneumatic Suspension   60
Air Brakes   62
Self-Contained Underwater Breathing Apparatus (SCUBA)   66
Self-Contained Breathing Apparatus   68
Questions   70

Chapter 3. Compressor Types   71

Squirrel Cage Blowers   73
Centrifugal Blowers   74
Regenerative Blowers   75
Rotary Lobe Blowers (Roots)   77
Liquid Ring Compressors   77
Diaphragm Compressor   78
Piston Compressor   79
Double-Acting Compressors   80
Two-Stage Compressors   83
Screw Compressor   85
Rotary Vane Compressors   88
Turbo Compressors   90
Axial Compressors   91
Injector Compressors   91
Questions   93

Chapter 4. The Conservation of Mass and Energy as Applied to
Compressed Air   95
Questions   98

Chapter 5. The Compressed Air Supply   99

Screw Compressors   99
Duplex Compressors   101
Redundant Compressors   103
Duplex Controllers   103
After Coolers   104
Air-Cooled After Coolers   106
Water-Cooled After Coolers   107
Air Dryers   108
Instrument Air   110
Field Compressors   113
Questions   116

Chapter 6. Compressed Air Dryers   117

Traps   118

Plate Separators   124
Coalescing Filters   125
Water-Cooled Aftercoolers   128
CFC-Based Refrigerated Dryers   128
Elliott Cycle Refrigerated Dryers   135
Standard Installations   141
Duplex Installations   142
Applications Greater Than 120 SCFM   144
Instrumentation Air Systems   144
Saltwater Prepared Dryers   145
Ice Bath Dryers   146
Desiccant Dryers   150
Deliquescent Dryers   154
Membrane Dryers   154
Absorption Dryers   157
Dryer Performance, Types, and Applications   157
Oil/Water Separators   158
Questions   161

Chapter 7. Support Components   163

Filters   164
Pressure Regulators   165
Lubricators   166
Filter, Regulators, and Lubricators Guide   167
Safety Relief Valves   168
Pressure Relief Valves   169
Check Valves   170
Expansion Joints and Vibration Isolators   172
Hoses and Hose Connectors   173
Retractable Hoses   174
Hose Clamps   175
Hose Connectors   176
Drain Valves   178
Commercial Pipe Fittings   181
Compression Fittings   183
Pipe Hangers   184
Ball Valves   186
Gate Valves   186
Globe Valves   189
Butterfly Valves   189
Solenoid Valves   191
Pilot Valves   192
Needle Valves   193
Speed Control Valves   194
Tire Valve Fittings   195
Mufflers   195
Intake Air Filters   195

Pressure Gauges   196
Pressure Switches   198
Compressor Control Switches   198
Continuous-Run Unloader Valves   199
Vee Belts   200
Sound Dampening Enclosures   201
Compressor Mounts   202
Skid Mounting   203
Questions   209

Chapter 8. Pneumatic Controls   211

In-House Pneumatic Symbols   217
Questions   221

Chapter 9. Electrical Controls   223

Single-Pole Control   225
Double-Pole Control   226
Motor Starters   227
120-VAC Control Loops   228
Sensor Loops   229
On, Off, and Automatic Functions   231
Sensors   232
Delta/Wye Motor Starters   233
Soft Starters   235
Variable Frequency Drives   237
Programmable Logic Controllers   238
Microprocessor-Based Controllers   240
Emergency Stop Circuit   240
Power Disconnects   243
Service Outlets   243
Lockout/Tagout Programs   243
Duplex Controllers   246
Parallel Duplex Operation   249
Weekly and Daily Duplex Operation   250
Wire Types   251
Din Rails and Wire Guides   252
Questions   254

Chapter 10. Maintenance   255

Supply Side   256
Distribution   259
Applications   261
Maintenance Program   262
Replacement Parts Inventory   264
Questions   267

Chapter 11. Energy and Costs Associated with Compressed Air   269

Leaks   270
Misuse   272
The Compressed Air Supply   272
Heat   272
The Motor   273
Drive Belts   274
Master Regulator   274
Water   274
Water Damage   274
Air Versus Electricity   277
Questions   279

Chapter 12. System Models   281

Contractor Compressors   281
Automotive Repair   283
Small Manufacturing Plants   288
Dry-Cleaning Operations   291
Heating, Ventilation and Air Conditioning (HVAC)   293
Dental Systems   296
Home Shop   298
Medium to Large Manufacturing Facilities   298
Instrument Air   302
High Purity   305
Service Trucks   310
Field and Construction   312
SCUBA and SCBA Charging Systems   315
Questions   320

Chapter 13. Specifying a Compression System   321

Estimating Required SCFM   322
Selecting a Compressor   323
Master Regulator   324
Receiver Size   325
Pipe Sizing   326
Remote Receivers   327
Cost Savings   329
Miscellaneous   330
Distribution Layouts   331
Plumbing   333
Buried Pipe   334
Plumbing Tools   336
Questions   339

Chapter 14. Conducting an Air Audit   341

Distribution Map   343
SCFM Requirements   343
Leaks   343
Hoses   345
Filters   345
Regulators   345
Lubricators   346
Remote Receivers   346
Master Regulator   347
Primary Receiver   347
Dryers   348
Compressors   348
Routine Maintenance   349
Employee Training   349
Sample Outline for Conducting an Air Audit   350
Questions   353

Chapter 15. Single-Stage Home-Built Air Compressor   355

Engine Preparation Instructions   357
Selecting the Sheaves (Items 21 and 22)   361
Assembly Instructions   362
Some Notes and Comments   363

Chapter 16. Home-Built Compressed Air Drying System   365

Appendix A. Glossary of Terms and Abbreviations   369

Appendix B. Miscellaneous Technical Information   371

Appendix C. Occupational Safety & Health Administration (OSHA)   397
Laws and Regulations Concerning Compressed Air

Appendix D. Answers to Chapter Questions   401

Index   403

List of Figures and Tables

Chapter 1

Figure 1-1	Fireplace bellows.	2
Figure 1-2	Manual powered diving compressor.	3
Figure 1-3	Bicycle tire pump.	5
Figure 1-4	Five HP compressor, circa 1917.	6
Figure 1-5	Contractor compressor.	7

Chapter 2

Figure 2-1	Die grinder.	10
Figure 2-2a	Compact blow-off gun.	10
Figure 2-2b	Blow-off gun.	11
Figure 2-3	Safety blow-off gun.	11
Figure 2-4	Simple air chuck.	12
Figure 2-5	Tire gauge.	13
Figure 2-6	Tire inflator. (*Milton Industries, Inc.*)	13
Figure 2-7	High accuracy tire gauge.	13
Figure 2-8	Portable air tank.	14
Figure 2-9	Air pressure fire extinguisher.	15
Figure 2-10	Air cylinder.	16
Figure 2-11	Diaphragm cylinder.	17
Figure 2-12	Double-acting cylinder.	18
Figure 2-13	Bolt-together cylinder.	18
Figure 2-14	Spring return cylinder.	19
Figure 2-15	Double-ended cylinder.	19
Figure 2-16	Nose and pivot mounts.	20
Figure 2-17	Diaphragm cylinder.	20
Figure 2-18	Electrical cylinder control.	21
Figure 2-19	Manual cylinder control.	21
Figure 2-20	Spring return cylinder control.	22
Figure 2-21	Force generated from cylinders.	23
Figure 2-22	Piston diameter.	24
Figure 2-23	Cylinder installation.	25

Figure 2-24   Hydraulic pressure amplifier.   25
Figure 2-25   Hydraulic clamping system.   26
Figure 2-26   Pneumatic/hydraulic pump.   27
Figure 2-27   Needle scaler.   28
Figure 2-28   Air hammer set.   28
Figure 2-29   Jack hammer.   30
Figure 2-30   Rivet/bolt cutter.   31
Figure 2-31   Rotary vane motor.   32
Figure 2-32   Air motor.   32
Figure 2-33   Die grinder.   33
Figure 2-34   Air ratchet.   34
Figure 2-35   Hand drill.   34
Figure 2-36   Disk sander.   34
Figure 2-37   RPM reduction drive.   35
Figure 2-38   Heavy duty hand drill.   35
Figure 2-39   Impact wrench mechanism.   36
Figure 2-40   Air wrench.   37
Figure 2-41   High speed dental drill (hand piece).   38
Figure 2-42   Air brush.   38
Figure 2-43   Paint gun.   39
Figure 2-44   Gravity feed spray gun.   40
Figure 2-45   Production paint tank.   41
Figure 2-46   Siphon gun.   41
Figure 2-47   Sandblaster schematic.   43
Figure 2-48   Sand-blasting cabinet.   44
Figure 2-49   Tumble blaster.   45
Figure 2-50   Air nailer.   46
Figure 2-51   Air stapler.   47
Figure 2-52   Box stapler. (*ISM Carton Closing Company, Inc.*)   47
Figure 2-53   Pneumatic caulk gun.   48
Figure 2-54   Pneumatic grease gun.   49
Figure 2-55   Pneumatic pop riveter.   50
Figure 2-56   Duel diaphragm pump.   51
Figure 2-57   Vacuum generator.   52
Figure 2-58   Air bearing.   53
Figure 2-59   Spray mister.   54
Figure 2-60   Rotary ball vibrator.   54
Figure 2-61   Piston vibrator.   55
Figure 2-62   Automobile lift.   56
Figure 2-63   Air-over-hydraulic bottle jack schematic.   57
Figure 2-64   Air-over-hydraulic bottle jack.   57
Figure 2-65   Barrel jack.   58
Figure 2-66   Pneumatic floor jack.   59
Figure 2-67   Bag jack.   60
Figure 2-68   Leaf spring suspension with auxiliary air bag.   61
Figure 2-69   Air ride suspension.   61
Figure 2-70   Air brake.   62
Figure 2-71   Basic air brake system.   63

Figure 2-72   Train car air brakes.   64
Figure 2-73   Train air brake schematic.   65
Figure 2-74   SCUBA system schematic.   66
Figure 2-75   Self-contained underwater   67
              breathing apparatus (SCUBA).
Figure 2-76   SCBA system schematic.   68
Figure 2-77   Self-contained breathing apparatus (SCBA).   69

Chapter 3

Figure 3-1   Bicycle tire pump.   72
Figure 3-2   Single cylinder piston pump.   73
Figure 3-3   Squirrel cage blower.   74
Figure 3-4   Centrifugal blower.   75
Figure 3-5   Regenerative compressor.   76
Figure 3-6   Regenerative compressor.   76
Figure 3-7   Two lobe (roots) compressor.   77
Figure 3-8   Liquid ring compressor.   78
Figure 3-9   Diaphragm compressor cycle.   79
Figure 3-10   Diaphragm compressor.   79
Figure 3-11   Single-stage piston compressor cycle.   80
Figure 3-12   Single cylinder compressor.   81
Figure 3-13   Double-acting compressor schematic.   82
Figure 3-14   Single-stage double-acting compressor.   83
Figure 3-15   Two-stage piston compressor cycle.   84
Figure 3-16   Two-stage reciprocating compressor.   85
Figure 3-17   Two-stage pump.   86
Figure 3-18   Screw compressor.   86
Figure 3-19   Motor/air end assembly.   87
Figure 3-20   Rotary vane compressor.   88
Figure 3-21   Rotary vane dental compressor.   89
Figure 3-22   Turbo compressor.   90
Figure 3-23   Axial compressor.   91
Figure 3-24   Injector compressor.   92

Chapter 4

Figure 4-1   Expansion/compression graphic.   96

Chapter 5

Figure 5-1   Reciprocating compressor schematic.   100
Figure 5-2   Screw compressor system.   100
Figure 5-3   Duplex compressor.   101
Figure 5-4   Duplex installation w/two vertical compressors.   102
Figure 5-5   Duplex installation for peak load applications.   102
Figure 5-6   Redundant screw compressor layout.   103

Figure 5-7   24 h/7 day duplex controller.   104
Figure 5-8   Redundant screw system.   105
Figure 5-9   Peak demand system settings.   105
Figure 5-10 Air-cooled after cooler.   106
Figure 5-11 Reciprocating compressor with after cooler.   107
Figure 5-12 Stand-alone after cooler.   108
Figure 5-13 Stand-alone after cooler for large compression systems.   109
Figure 5-14 Water-cooled after cooler schematic.   109
Figure 5-15 Water-cooled after cooler.   110
Figure 5-16 Packaged reciprocating compressor w/CFC dryer.   110
Figure 5-17 Screw compressor w/CFC dryer.   111
Figure 5-18 Instrument air system.   112
Figure 5-19 Improved instrument air system.   114
Figure 5-20 Gasoline-powered field compressor.   115
Figure 5-21 Diesel-powered screw compressor.   115

Chapter 6

Figure 6-1   Are there times when it feels like your air   118
                 tools are operating off of a garden hose?
Figure 6-2   Drop trap.   119
Figure 6-3   Welded drop trap.   119
Figure 6-4   Drop trap construction with threaded pipe fittings.   120
Figure 6-5   Automatic electronic drain.   121
Figure 6-6   Line trap schematic.   121
Figure 6-7   Commercial line trap.   122
Figure 6-8   Cyclone trap.   123
Figure 6-9   Commercial cyclone trap.   123
Figure 6-10 Plate separator.   124
Figure 6-11 Commercial plate separator.   125
Figure 6-12 Micro grid coalescing filter.   126
Figure 6-13 Stacked plate element.   126
Figure 6-14 Porous media element.   127
Figure 6-15 Coalescing filter installation.   127
Figure 6-16 Micro grid in a systems role.   128
Figure 6-17 Water-cooled aftercooler.   129
Figure 6-18 CFC refrigerated dryer.   129
Figure 6-19 Reheated output.   131
Figure 6-20 High-temperature input.   131
Figure 6-21 Commercial refrigerated dryer.   132
Figure 6-22 Cycling refrigerated air dryer.   133
Figure 6-23 High-capacity cycling dryer.   134
Figure 6-24 Packaged compressor with/CFC dryer.   135
Figure 6-25 Screw compressor with/CFC dryer.   136
Figure 6-26 Elliott refrigeration cycle.   136
Figure 6-27 Elliott cycle refrigerator. (*Courtesy of Air Options Inc.*)   137
Figure 6-28 Elliott cycle dryer sectional view.   139
                 (*Courtesy of Air Options Inc.*)

Figure 6-29   2000 SCFM Elliott cycle dryer.   140
(Courtesy of Air Options Inc.)

Figure 6-30   Elliott cycle dryer/compressor schematic.   140
(Courtesy of Air Options Inc.)

Figure 6-31   Elliott cycle dryer, vertical compressor.   141
(Courtesy of Air Options Inc.)

Figure 6-32   Elliott cycle dryer, horizontal compressor.   142
(Courtesy of Air Options Inc.)

Figure 6-33   Elliott cycle dryer, base mounted compressor.   143
(Courtesy of Air Options Inc.)

Figure 6-34   Elliott cycle dryer, duplex compressor.   143
(Courtesy of Air Options Inc.)

Figure 6-35   Elliott cycle dryer, duplex system.   145
(Courtesy of Air Options Inc.)

Figure 6-36   Elliott cycle dryer, instrument air system.   146
(Courtesy of Air Options Inc.)

Figure 6-37   Salt water prepared dryer.   147
(Courtesy of Air Options Inc.)

Figure 6-38   Ice bath dryer schematic.   148
(Courtesy of Air Options Inc.)

Figure 6-39   20 SCFM ice bath dryer.   149
(Courtesy of Air Options Inc.)

Figure 6-40   Ice bath dryer and compressor.   149
(Courtesy of Air Options Inc.)

Figure 6-41   Desiccant dryer schematic.   150

Figure 6-42   Canister dryer.   151

Figure 6-43   Desiccant dryer at application site.   152

Figure 6-44   Twin tower desiccant dryer.   153

Figure 6-45   Low dew point system.   154

Figure 6-46   Deliquescent dryer schematic.   155

Figure 6-47   Deliquescent dryer.   155

Figure 6-48   Membrane dryer schematic.   156

Figure 6-49   Commercial Membrane Dryer.   156

Figure 6-50   Absorption dryer.   157

Figure 6-51   Dew point scale.   158

Figure 6-52   Oil/water separator. *(Courtesy of Air Options Inc.)*   159

Figure 6-53   Commercial oil/water separator system.   160
(Courtesy of Air Options Inc.)

Chapter 7

Figure 7-1   Various support components.   163

Figure 7-2   Particulate filter housing and pleated filter element.   164

Figure 7-3   Master regulator.   165

Figure 7-4   Pressure regulator.   166

Figure 7-5   Lubricator.   167

Figure 7-6   Filter, Regulators and Lubricators guide.   168

Figure 7-7   Safety relief valve.   169
Figure 7-8   Adjustable pressure relief valve.   170
Figure 7-9   In-tank check valve.   171
                 (Courtesy of Control Devices, Inc.)
Figure 7-10 In-line check valve.   171
Figure 7-11 Flexible metal hose.   172
Figure 7-12 Tube loop.   172
Figure 7-13 Standard compressed air hose.   173
Figure 7-14 Hose reel.   174
Figure 7-15 Coil hose.   175
Figure 7-16 Hose clamps.   176
Figure 7-17 Hose fittings.   177
Figure 7-18 SNPT screw connector.   177
Figure 7-19 Industrial quick disconnect.   177
Figure 7-20 Universal hose coupling.   177
Figure 7-21 Universal hose connector manifold.   178
Figure 7-22 Electronic drain.   179
Figure 7-23 Float drain.   179
Figure 7-24 External seat drain cock.   180
Figure 7-25 Plug valve.   180
Figure 7-26 Cable drain valve.   181
Figure 7-27 "Y" strainer.   181
Figure 7-28 Strainer.   182
Figure 7-29 Commercial pipe fittings.   183
Figure 7-30 Compression fitting.   184
Figure 7-31 Commercial brass compression fittings.   184
Figure 7-32 Saddle type pipe hanger.   185
Figure 7-33 Ball valve schematic.   187
Figure 7-34 Commercial ball valve.   187
Figure 7-35 Three-way ball valve.   187
Figure 7-36 Four-way ball valve.   188
Figure 7-37 Gate valve schematic.   188
Figure 7-38 Gate valve.   189
Figure 7-39 Globe valve schematic.   190
Figure 7-40 Globe valve.   190
Figure 7-41 Butterfly valve schematic.   191
Figure 7-42 Commercial butterfly valve.   192
Figure 7-43 Solenoid valve.   192
Figure 7-44 Pilot valve.   193
Figure 7-45 Needle valve schematic.   193
Figure 7-46 Needle valve.   194
Figure 7-47 Speed control valve.   194
Figure 7-48 Tire valve fitting.   195
Figure 7-49 Pneumatic muffler.   196
Figure 7-50 Intake air filter.   196
Figure 7-51 Bottom mount pressure gauge.   197
Figure 7-52 Rear mount pressure gauge.   197
Figure 7-53 Gauge internals.   198

Figure 7-54 Pressure switch.   198
Figure 7-55 Compressor control switch.   199
Figure 7-56 Continuous-run unloader valve.   200
                 (*Courtesy of Control Devices, Inc.*)
Figure 7-57 Determining belt length.   200
Figure 7-58 Common commercial v-belt sizes.   201
Figure 7-59 Sound damping enclosure.   202
Figure 7-60 Compressor mount.   203
Figure 7-61 Skid mounted compression system.   204
Figure 7-62 Skid components.   205
Figure 7-63 Mount pad.   205
Figure 7-64 Fluid lip detail.   206
Figure 7-65 Drain methods.   206
Figure 7-66 Sample skid engineering drawing.   207
Figure 7-67 Incorrect expansion connection.   208
Figure 7-68 Plumbed to allow for expansion.   208

Chapter 8

Figure 8-1   Pressure regulator.   215
Figure 8-2   Workstation.   215
Figure 8-3   Schematic representation.   215
Figure 8-4   Cylinder control system.   216
Figure 8-5   Schematic representation.   216
Figure 8-6   Compression system.   217
Figure 8-7   Schematic representation.   217
Figure 8-8   Cylinder installation.   220
Figure 8-9   Schematic representation with standard symbols.   220
Figure 8-10 Typical in-house schematic with legend.   220
Figure 8-11 Schematic for Chap. 8 questions.   221

Table 8-1   Pneumatic Symbols   212
Table 8-2   In-House Pneumatic Symbols   219

Chapter 9

Figure 9-1   120, 240, 480 VAC 60 Hz Graphic.   224
Figure 9-2   Single-pole control circuit.   225
Figure 9-3   Compressor with manual/automatic pressure switch.   225
Figure 9-4   Double-pole control circuit.   226
Figure 9-5   Compressor with double-pole pressure switch.   226
Figure 9-6   Three-phase motor controller.   227
Figure 9-7   Compressor with motor controller.   227
Figure 9-8   Motor controller with 120-VAC control circuit.   228
Figure 9-9   Commercial motor controller with   229
                 120-VAC control circuit.
Figure 9-10 Motor controller with sensor loop.   230

Figure 9-11   Commercial motor controller with sensor loop.   230
Figure 9-12   Motor control circuit with run/auto mode.   231
Figure 9-13   Motor controller with run/auto mode.   232
Figure 9-14   Various sensors.   233
Figure 9-15   Delta/Wye motor controller schematic.   234
Figure 9-16   Delta/Wye motor controller.   235
Figure 9-17   Commercial soft starter.   236
Figure 9-18   Switching cycle.   236
Figure 9-19   Commercial variable frequency drive.   237
Figure 9-20   Variable frequency soft start cycle.   238
Figure 9-21   Controller with PLC.   239
               (*Courtesy of Automationdirect, Inc.*)
Figure 9-22   Microprocessor-based compressor controller.   241
Figure 9-23   Microprocessor/variable frequency drive.   242
Figure 9-24   Emergency stop schematic.   242
Figure 9-25   Power disconnect.   244
Figure 9-26   Auxiliary AC outlet installation.   245
Figure 9-27   Lockout safety items.   246
Figure 9-28   Power disconnect with lockout tags.   247
Figure 9-29   Two separate compressors set up with   248
               a duplex controller.
Figure 9-30   Duplex controller schematic.   248
Figure 9-31   Parallel duplex control schematic.   249
Figure 9-32   24 h/7 day toggle controller.   250
Figure 9-33   24 h/7 day toggle controller schematic.   251
Figure 9-34   Standard wire sizes.   251
Figure 9-35   Type MTW wire.   252
Figure 9-36   DIN rail.   252
Figure 9-37   Wire guide.   253

Chapter 10

Figure 10-1   Valve/universal hose coupling.   256
Figure 10-2   Maintenance items.   258
Figure 10-3   Stand-alone compressor room.   260
Figure 10-4   Bolted gate valve.   260
Figure 10-5   Bolted cylinder.   262

Table 10-1   Supply Maintenance Items   263
Table 10-2   Distribution Maintenance Items   264
Table 10-3   Application Maintenance Items   265

Chapter 11

Figure 11-1   Energy saving items within the distribution system.   271
Figure 11-2   Energy savings items at the supply.   275

Figure 11-3 Air motor performance deterioration due to   276
water contamination.
Figure 11-4 Production variation between electric and   278
pneumatic grinders due to fatigue.
Figure 11-5 Percentage of actual production time during period.   278

Chapter 12

Figure 12-1   Contractor compressor.   283
Figure 12-2   Two-zone distribution system for new   284
home construction.
Figure 12-3   Two-stage compressor with dryer.   285
Figure 12-4   Auto repair shop workstation.   286
Figure 12-5   Paint booth workstation.   287
Figure 12-6   Four bay paint and body shop distribution system.   288
Figure 12-7   Small manufacturing plant compressed air supply.   289
Figure 12-8   Plant floor workstation.   290
Figure 12-9   Cylinder control station.   291
Figure 12-10 Spine distribution system for small   292
manufacturing plant.
Figure 12-11 Single-stage compressor with dryer.   293
Figure 12-12 Pneumatic equipment feed station.   294
Figure 12-13 Airbrush workstation.   295
Figure 12-14 Distribution system for a neighborhood   296
dry cleaning plant.
Figure 12-15 HVAC duplex compressor and dryer.   296
Figure 12-16 HVAC control pressure port.   297
Figure 12-17 Dental compression system.   298
Figure 12-18 Home shop compression system.   299
Figure 12-19 Redundant system for medium manufacturing plant.   300
Figure 12-20 Plant floor workstation.   301
Figure 12-21 Overhead isolation valve and line plant.   302
Figure 12-22 Loop distribution system.   303
Figure 12-23 Utility and instrument air system.   304
Figure 12-24 Chemical plant utility trunk.   306
Figure 12-25 Chemical plant primary distribution system.   307
Figure 12-26 High purity compressed air system.   308
Figure 12-27 High purity workstation.   309
Figure 12-28 Compressed air quality analytical instrument.   310
Figure 12-29 Service truck compression system.   311
Figure 12-30 Various tools required for tractor-trailer tire service. 312
Figure 12-31 Trailer mounted diesel-powered screw compressor.   313
Figure 12-32 Field distribution components.   314
Figure 12-33 2000 SCFM field dryer.   314
Figure 12-34 Field distribution layout.   315
Figure 12-35 Four-stage 5000 psi breathing air compressor.   315
Figure 12-36 SCUBA compressed air supply.   317
Figure 12-37 SCUBA tank filling station.   318

Chapter 13

Figure 13-1	Two-stage reciprocating compressor.	324
Figure 13-2	Master regulator.	325
Figure 13-3	Commercial receivers.	327
Figure 13-4	Spine distribution system.	331
Figure 13-5	Loop distribution system.	332
Figure 13-6	Plant plumbing.	333
Figure 13-7	Secondary runs and remote receivers.	335
Figure 13-8	Underground piping.	335
Figure 13-9	Basic plumbing tools.	336
Figure 13-10	Pipe vise.	337
Figure 13-11	Electric pipe threading machine.	338
Table 13-1	SCFM Work Sheet	323
Table 13-2	Receiver Size Chart	326
Table 13-3	Recommended Pipe Sizes	328

Chapter 14

Figure 14-1	System map.	344
Figure 14-2	Poorly maintained compressor graphic.	350

Chapter 15

Figure 15-1	Home-built compressor.	358
Figure 15-2	Base.	359
Figure 15-3	Tank assembly.	359
Figure 15-4	Mount bracket.	360
Figure 15-5	Electrical system.	360
Figure 15-6	Pick-up tube.	360

Chapter 16

Figure 16-1	Ice bath dryer.	366
Figure 16-2	Trap assembly.	366
Figure 16-3	Ice bath dryer/compressor setup.	367
Figure 16-4	Dry ice dryer.	368
Figure 16-5	Dry ice dryer/ice bath dryer/compressor setup.	368

Preface

The purpose of this book is to provide a basic understanding of the compressed air systems, equipment, and technologies that are available on the market today. You might have many years of experience with a specific compression system, but the catalogs and sales literature just aren't providing you with the information you need. Maybe you've just been assigned the compression system as part of your responsibilities and you have no experience whatsoever with this type of equipment. Oftentimes, the compressors are the bastard child of a company's capital equipment and no one in particular is responsible for them. If you head up the company's maintenance department, the compressors probably fall under your responsibilities. You may be a salesman who works with pneumatic-powered equipment and, from time to time, are asked to comment on your customer's compression systems. In any case, this book will give you the baseline information that will allow you to make well-informed decisions about specifying, expanding, or assessing a compression system.

Like many technical people, I felt that I had a clear conceptual understanding of air compression, I mean, what's to know? You suck in some air, you smash it into a tank. Simple! I even built an air compressor from an old auto AC pump and a piece of pipe when I was in high school.

In the early days of my career whenever I had to deal with air compression, I was bombarded with a myriad of information from all sorts of manufacturers and dealers, which, in most cases, seemed contradictory. I was forced to rely on the advice and the expertise of sales people and factory technical support, whose primary job was to sell their company products. Like almost anyone that is placed into that position, I generally selected the cheapest solution that was presented to me and in doing so, my employers and I slowly got a very costly education in air compression.

As it turns out, buying the least expensive alternative is rarely the right decision in the world of compressed air. When using price as your primary criteria, you generally wind up with a system that is inadequate for your needs and, invariably, are forced to spend much more money down the road than if you had just purchased the right equipment in the first place. I can tell you from actual experience, your boss doesn't want to hear "We need to buy a second air compressor, because the one we just got isn't big enough." This can really put you in a spot. Similarly, if you're a salesperson, you really don't want your customer calling you up 3 months after you sold him a system and hear "The motor on the compressor burned up this morning. What's the deal? You said that this compressor was big enough."

I'm not suggesting that you need to be a compression engineer to properly select an air compressor. However, a cursory understanding of air compression and the basics of the equipment will arm you with enough information so that you can ask the right questions. This will allow you to sift through all the information that you will receive and make reasonably informed decisions for your specific application. This will also have the added advantage of keeping you from looking completely incompetent in the eyes of your boss or customers.

Acknowledgments

Special thanks go to Rozanne Hill for her support in getting me through the effort of writing this book. Her input was invaluable. I don't suppose I could have finished without her help.

I would also like to thank Kevin Marchant for lending his expertise in the area of electrical controls.

Special thanks also go to Jeff Edwards and Timothy Elliott for helping me edit the book by lending their technical expertise and suggestions.

I'll know I have achieved literary greatness on the day I am able to purchase a copy of one of my books in a garage sale for a dollar.

The History of Compressed Air

Everyone has experienced compressed air, though most people do not realize it. Our lungs are a low-pressure air compressor. In order to breathe, we expand our lungs, creating a negative internal pressure, which draws in outside air. To expel the air, we contract our lungs, which creates a positive pressure and the air flows out. Neanderthal man used his lungs as a compressor to blow air on a glowing ember in order to build up his fires. Tribes in Africa and South America used their lungs as a compressed air source to power their blowguns. Every time you speak or whistle you use compressed air to create sound. When you blow up a balloon, compressed air is stored as an energy form inside the balloon every time you exhale. After the balloon is fully inflated, it can be set aside and the compressed air is available for release whenever you like. You and the balloon represent a simple air compression system.

The earliest type of mechanical compressor was the hand-operated bellows. Historians estimate that these devices were being used as early as 3000 BC to support rudimentary metal smelting. As metallurgy became more advanced, bellows were built larger and made more efficient with the advent of foot power. Eventually, large water powered bellows were built that were capable of feeding rather large forging operations. Figure 1-1 shows a typical set of fireplace bellows. These units can still be purchased today to make fire building much easier. Starting a fire with wet logs is no picnic; however, stoking the kindling with a set of bellows produces a substantially hotter flame and the wet logs start much quicker. Another early use for bellows was using the air jet to blow ashes out of the fireplace during periodic cleanings.

Large foot-and-water-operated bellows can be seen in action in many of the historical blacksmith shops that dot the country. It is quite fascinating to watch these early compressors in action and the effect that

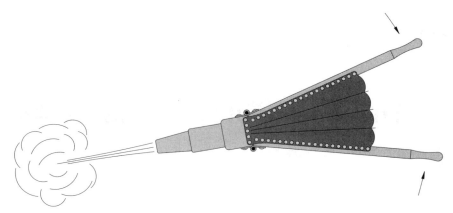

Figure 1-1 Fireplace bellows.

they have on the forge temperature. As the industrial revolution progressed, many bellows came to be replaced by hand-driven centrifugal blowers. These units are essentially a turbo charger with a crank, gearbox, and flywheel mounted in a position that the blacksmith can conveniently access.

In 1650, German physicist Otto Von Guericke invented what is recognized as the first air pump. He used his device to conduct studies on pressure, vacuum, combustion, and respiration. By the late 1700s, blowers were in common use supplying forced air to metal furnaces and in the ventilation of large mining operations. By the early 1800s, engineers were considering the use of compressed air as a viable energy source. Large industrial plants of the time relied heavily on steam for their power requirements. Steam, however, has its limits and is less than ideal for power transmission over long distances. Compressed air, on the other hand, seemed to be an ideal solution for the distance problem and distributed compressed air systems were born.

Another use for compressed air, which came about in the early 1800s, was for underwater diving. Until the development of compression systems, which could produce reasonable pressures, a diver's downtime was limited to his lung capacities. This made salvage work virtually impossible in all but shallow waters. When high-pressure pumps were developed, this changed. Early dive suits were little more than a helmet, which was strapped to the diver's head. The helmet acted as a small diving bell. Compressed air was pumped down from the surface compressor and into the helmet. Excess compressed air simply escaped out from the bottom of the helmet and bubbled up to the surface. In this manner, the head of the diver was at all times safely enclosed within an air pocket.

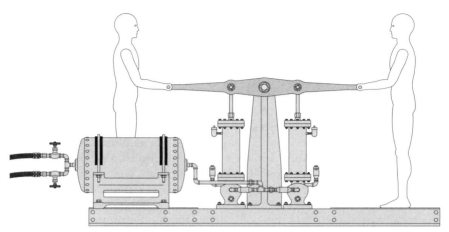

Figure 1-2   Manual powered diving compressor.

Early dive compressors were hand-operated devices that were usually mounted on the deck of a ship or barge. Figure 1-2 shows a hand-operated diving pump of the early 1800s. These pumps were usually powered by four hardy men, but even so they could not produce particularly high pressures or flow rates, so the depth at which a diver could work was limited. As sail power was replaced with steam, the human power of dive compressors was replaced as well, providing substantial improvements in pressures, flow rates, dive depths, and downtimes.

In the early part of the 1800s, compressed air was starting to be utilized in mining and tunneling operations to power lightweight rock drills and hammers. The speed at which compressed air allowed subterranean workers to move through rock was unprecedented. Air compressors could be set up at the surface and piping and hoses could be easily routed throughout the mine, providing a readily accessible power source.

By the mid-1800s, engineers were at odds with one another over whether electricity or compressed air should be the power source of the future. In 1888, an Austrian engineer named Viktor Pop installed a 1500-kW compression system in Paris. Existing sewers were used as a distribution system and Paris became the first city to offer compressed air as a public utility. By 1891, the Paris compressors had been increased in size to 18,000 kW. The compressed air was used to activate nearly 58,000 clocks and about 4000 elevators, as well as powering all sorts of industrial equipment.

In 1869, the U.S. Army Corps of Engineers and the New York City Council approved the construction of the Brooklyn Bridge. The tower foundations presented a difficult problem. The silt that made up the floor of the river would allow water to flow into standard caissons faster than the pumps could remove it. Special 3000-ton pressurized caissons were designed and placed into position. The water was forced out through the bottom of the caisson by applying compressed air to the sealed assembly. The workers had to move in and out of the caissons through a series of air locks. It was during this construction that the earliest cases of the bends were experienced, referred to at the time as caisson disease. During the course of the project, 20 men lost their lives to caisson disease and the chief engineer, Washington Robeling, was paralyzed.

In 1845, Robert Thompson invented the pneumatic tire and was granted a patent on the idea. However, his design was costly and never became a success. In 1888, John Dunlop was granted a patent for the first practical pneumatic tire and his design was a great success leading to the tire industry of today. Along with the pneumatic tire came flat tires and the need to pump them back up. The hand-operated tire pump is most likely the first compressor designed for a mass market. By the time pneumatic tires became standard equipment on bicycles, cars, and trucks, there were millions of hand-operated pumps under seats and in trunks, garages, and gas stations. Figure 1-3 shows a typical hand-operated tire pump.

During the later part of the 1800s, compressed air was limited to rather large companies and operations. This is because the pumps required a power source, which was usually in the form of a steam boiler and engine. Smaller businesses could not afford a power source as expensive as a steam system, so they were forced to do without it. The advent of electrical power and compact electric motors transformed the industry in early 1900s. Manufacturing equipment no longer relied on a central steam engine for power. Instead, the equipment had a compact electric motor, and power was delivered in the form of electricity. Air compressors were also transformed by this development. In the early 1900s, several companies started to manufacture small, inexpensive air compressors that were aimed specifically at small business. These units allowed companies that previously could not afford compressed air to add a powerful new utility to their facilities. The compressed air industry grew steadily during the first half of the 1900s. Figure 1-4 shows a 5-hp, 2-stage compressor manufactured in 1917 for small businesses. These units did not differ very much from the compressors of today. They typically came with a pump, electric motor, and receiver as a complete packaged unit.

Along with low-cost air compressors came low-cost pneumatic tools and devices. Small businesses really could not take advantage of their new utility if they could not afford air tools. Manufacturers started to introduce low-cost, personal sized air tools for a wide variety of applications. Compressed air quickly became popular in manufacturing and

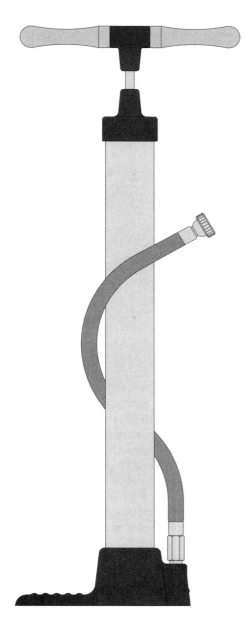

Figure 1-3 Bicycle tire pump.

service facilities throughout the country. World War II brought about an expansion of compressed air use. During the war anything that would increase production was quickly embraced and compressed air was no exception.

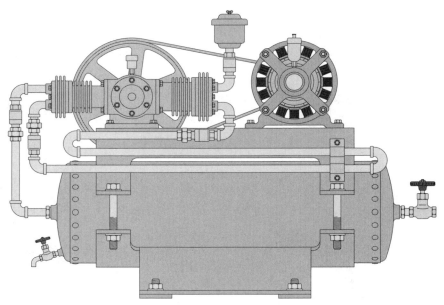

Figure 1-4 Five HP compressor, circa 1917.

In the 1970s, a new revolution in compressed air started to gain momentum with the advent of the home or contractor compressor (Fig. 1-5). These are small, ultra-low-cost compressors aimed at the individual use market. There are hundreds of dirt-cheap compressors in the market to choose from. These little compressors are found everywhere. Go to a new home construction site and you'll find several. There's a good chance that the neighbor who's always working on cars has one in the back of his garage. Walk into any pawnshop and see how many of these compact units are on the floor. Along with the cheap compressors came an explosion of inexpensive air tools. Nowadays it's uncommon to find an auto mechanic who doesn't have a good selection of air tools in his toolbox. Any home shop mechanic can afford an excellent compressed air system and all the tools that go along with it.

Well, the long and short of it is that compressed air has become the hidden utility. It has encroached on nearly every aspect of industry. Even professional buildings filled with lawyers have pneumatic heating, ventilation, and air conditioning (HVAC) controls and somewhere, hidden deep in the bowels of the building, is a compression system operating every single day. Remember the last trip you took to the amusement park? Well, very few of those rides could operate if it weren't for compressed air.

Figure 1-5 Contractor compressor.

The reason I refer to compressed air as the "hidden utility" is that the only area where it doesn't have an impact is the home. Millions of people go about their business every day without the slightest idea of how important compressed air is to our society. The closest they'll ever get to it is hearing the buzzing of air wrenches as they wait to get new tires on their car.

Questions*

1. What was the first type of mechanical compressor?
    (A) Lungs   (B) Blowgun   (C) Bellows   (D) Goatskins

2. What type of power was used to improve the bellows?
    (A) Foot   (B) Horse   (C) Water   (D) Rock

3. What type of blower replaced the bellows?
    (A) Wind   (B) Bags   (C) Piston   (D) Centrifugal

4. In which year was the first air pump invented?
    (A) 1776   (B) 1492   (C) 1650   (D) 3000 BC

5. How were early dive pumps powered?
    (A) Steam   (B) Human   (C) Horse   (D) Donkey

6. Name a type of utility that was available in Paris during the late 1800s.
    (A) Gas   (B) Water   (C) Compressed air   (D) Sewage

7. What type of caissons were used in the construction of the Brooklyn Bridge?
    (A) Standard   (B) Pressurized   (C) Wooden   (D) Rock

8. When was the first practical pneumatic tire patented?
    (A) 1888   (B) 1845   (C) 1650   (D) 2001

9. When did small compressors start to appear in the market?
    (A) 1800–1850   (B) 1900–1925   (C) 1940–1944
    (D) 1970–1975

10. What type of control do large HVAC systems use?
    (A) Electronic   (B) Flow   (C) Pneumatic   (D) Water

---

*Circle all that apply.

Compressed Air Applications

Uses for compressed air are as varied as for any other utility. These uses range from simply blowing dust off of a new cabinet to providing control pressure for entire petrochemical plants. Most of us have witnessed a busy auto repair shop and heard all the air tools buzzing away. Air tools provide a high power, light weight, safe and reliable alternative to other types of tools. From a control standpoint, compressed air represents a very powerful utility. Pneumatic controls are generally about one-third the size of their electrical counterparts and they have the added advantage of being impervious to water. Washdown rated electrical controls are extremely expensive and have a rather short life expectancy. Food processing plants, for instance, require regular washdown operations. Pneumatic controls are ideal for this industry and provide many years of service in these otherwise hostile environments. Another arena is petrochemical plants, where pneumatic controls shine. The bulk of most chemical plants are in the form of outside installations. Electrical controls cannot be exposed to the elements and require special construction or housings. Additionally, because of the flammable nature of the materials being handled, all electrical controls within a chemical plant must be explosion proof. Standard pneumatic controls, on the other hand, do not represent a fire or ignition hazard.

The auto repair industry relies heavily on pneumatic tools for every day operations. Most auto mechanics rate their air wrenches as some of their most important tools. If you stroll through a typical auto repair shop, you'll notice that each mechanic has two or three different air wrenches lying about his workstation.

Paint and body shops conduct most of their grinding and polishing operations with pneumatic tools. They use air wrenches to speed assembly and disassembly, and of course, apply all paint with pneumatic paint sprayers.

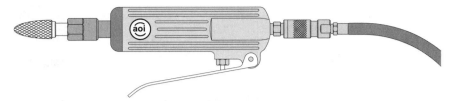

Figure 2-1 Die grinder.

Cabinet and furniture shops use pneumatic nailers and staplers almost exclusively. These businesses simply cannot compete without the speed and reliability that these tools provide their operation.

There are a myriad of air tools and pneumatic controls on the market today, and a detailed review of this subindustry is out of the scope of this text. Instead, this section intends to provide a brief overview of some of the more common air tools and pneumatic applications.

Blow-Off Guns

Blow-off guns are the most common use for compressed air. From blowing off chips in a machine shop to makeshift air brooms, these guns are extremely convenient tools. The blow-off gun (Fig. 2-2*a*) is simply a push button valve with a discharge nozzle to focus escaping air. Figure 2-2*b* shows a typical blow-off gun designed with a piston grip. This pattern of gun is preferred when operators are restricted to specific machines, such as lathes, mills, and punch presses. Figure 2-3 shows a safety blow-off gun that is intended to limit the discharge pressure to 30 psi.

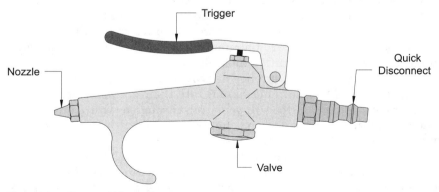

Figure 2-2a   Compact blow-off gun.

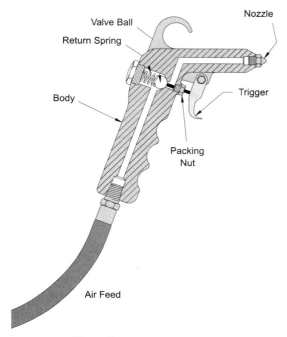

Valve Ball

Return Spring

Nozzle

Body

Trigger

Packing
Nut

Air Feed

Figure 2-2*b* Blow-off gun.

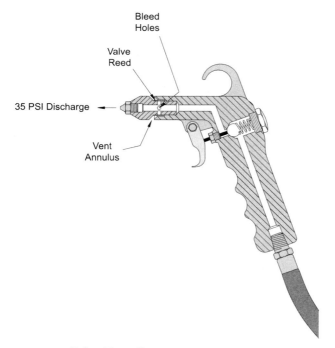

Bleed
Holes

Valve
Reed

35 PSI Discharge

Vent
Annulus

Figure 2-3 Safety blow-off gun.

This is accomplished by incorporating a high-pressure relief valve into the body of the gun. Safety blow-off guns are typically required by OSHA regulations. However, these guns are generally not favored by their operators because their low discharge pressure makes them less effective than their nonsafety counterparts.

Tire Inflators

Tire inflators are extremely convenient devices to have access to. The use of a tire inflator or "air chuck" is not limited to the auto repair shops. If you have a back dock, there should be an air chuck available. This is useful for airing-up the tires on delivery trucks as well as employee's cars. If your company is a large drive-on site, then place tire filling stations in various locations throughout the plant. If the filling stations are to be permanent, it's usually a good idea to plumb the air chucks without a quick disconnect. This helps to prevent the chucks from "walking away". If you have a compressor in your garage, you should have an air chuck. These little devices can make a big difference when you walk out in the morning and see a low tire on your wife's car.

Figure 2-4 shows a simple air chuck. This is a very inexpensive and handy tool to have squirreled away in your toolbox or desk. There are versions of the air chuck that have a locking mechanism, which allows the chuck to be locked to the tire valve. Caution should be exercised when using a locking air chuck. If the chuck is connected and the operator gets distracted the tire may be over pressurized, and in some cases a tire can rupture. In addition to the air chuck, it's a good idea to keep a tire gauge as well. A tire gauge, as shown in Fig. 2-5, should be a standard item in the glove box or your car. Figure 2-6 shows a typical tire inflator. These units are air chucks with a number of attributes to make regular use easier. A tire inflator will typically have a paddle valve, pressure gauge, and dual output air chucks mounted through a short section of hose. If you are a regular user of tire inflators, this is the unit of choice. These units are the preferred choice when installing permanent inflator stations.

Figure 2-7 shows a high-accuracy tire gauge, which can be constructed by using a 3-in., 100-psi gauge with 1-psi graduations. These types of

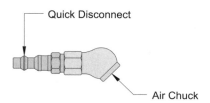

Quick Disconnect

Air Chuck

Figure 2-4 Simple air chuck.

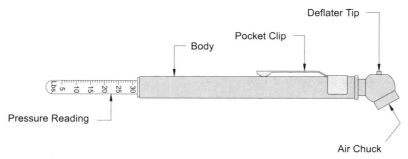

Figure 2-5 Tire gauge.

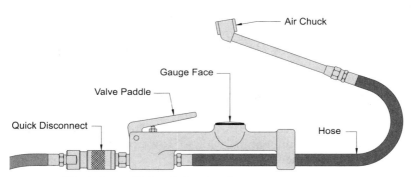

Figure 2-6 Tire inflator. (*Milton Industries, Inc.*)

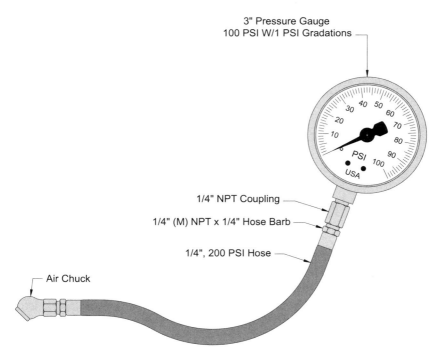

Figure 2-7 High accuracy tire gauge.

gauges are used by race car mechanics to adjust tire and shock absorber pressures. Small changes in these pressures can have a profound effect on the handling performance of the car.

Portable Air Tanks

For some field applications, most notably filling tires, a portable air tank or service tank can be utilized. These devices are little more than an air tank with a carrying handle, filler valve, and air chuck. The air chuck is typically connected to the tank with a 3- or 4-ft hose, which attaches to a tee. The tee normally has a pressure gauge, which shows the internal pressure of the tank. The filler valve is usually a tire-type valve, which allows the tank to be filled at any auto repair facility. These units are generally set up on four feet to provide a stable mount when in use. The tanks used normally have a 2- to 5-gal capacity because larger tanks are a little too bulky to be handled comfortably. Oftentimes, service tanks can be seen on the back of tow-trucks as they cruise down the road. If you have a trailer stored in a remote location, a service tank will allow you to conveniently air up the tires before moving it. Figure 2-8 shows a typical service tank.

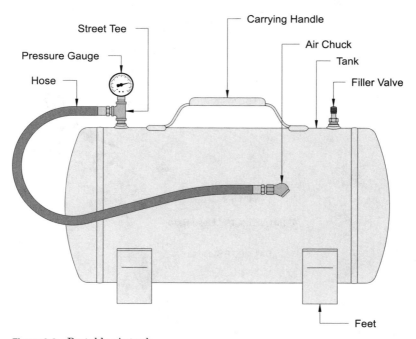

Figure 2-8   Portable air tank.

Fire Extinguishers

Another device that most of us are aware of, but probably wouldn't consider a pneumatic tool, is the pressurized air/water fire extinguisher (Fig. 2-9). These units are commonly found hanging in the halls of

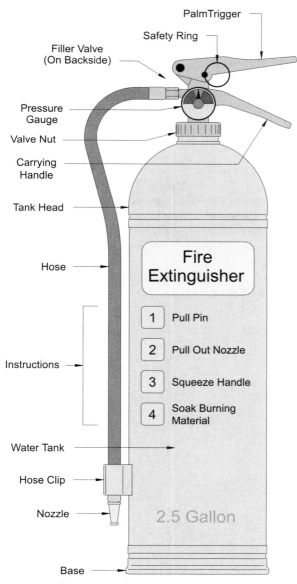

Figure 2-9   Air pressure fire extinguisher.

schools, hospitals, and office buildings. They rely heavily on compressed air for their operation. A valved cap is used to seal a small water tank. The valve assembly consists of a palm-actuated valve, safety ring, carrying handle, pressure gauge, hose with nozzle, and filler valve. After the tank is charged with water, usually two and a half gallons, the valve cap assembly is replaced on the tank and the remaining air space is pressurized with compressed air. The air is introduced through a tire filler tank valve that is generally opposite the pressure gauge. When pressurizing one of these extinguishers, the gauge should indicate in the "green" zone. If the gauge indicates anywhere in the red zones, the internal pressure should be adjusted back into the green. These types of fire extinguishers should not be used on electrical fires.

To operate the unit, simply pull out the safety ring, point the nozzle at the fire, and push down the palm trigger. The air charge forces the water out of the nozzle at a surprising rate.

Air Cylinders

Among the most common ways to take advantage of compressed air is through the use of cylinders. These devices offer a versatile and reliable way to convert the energy of compressed air into linear motion. By selecting different piston areas and adjusting the delivery pressure, a wide range of force can be realized.

There are two principal types of cylinders available in the market. The first (Fig. 2-10) is simply a cylinder with a piston and rod. These cylinders

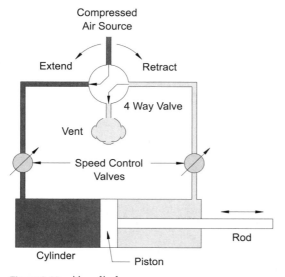

Figure 2-10   Air cylinder.

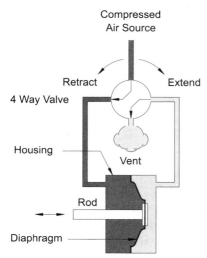

Figure 2-11   Diaphragm cylinder.

are available in a wide range of diameters and strokes. The second (Fig. 2-11) is the diaphragm actuator. These serve the same general purpose, except a diaphragm replaces the piston. These types of cylinders will generally endure water contamination better than their piston counterparts. They are normally used to produce high forces with low pressures by taking advantage of their large diaphragm area. The real drawback in diaphragm cylinders is their stroke length. Because the diaphragm cannot slide within a cylinder, there are practical limits that must be dealt with. The other drawback is that because their diaphragms cannot handle large differential pressures, these units are normally limited to relatively low pressures.

Cylinders are typically controlled with a venting four-way valve. When the valve is in one position, pressure is applied to one side of the piston and vented on the opposite side. These valves are available in many different configurations, which are applicable for almost any control situation.

Controlling the speed of the cylinder is another consideration in almost every application. Small, inexpensive speed control valves are available that can be adjusted to control the rate of either extension, contraction, or both.

The double acting cylinder (Fig. 2-12) is the most common cylinder type. These cylinders require air pressure in order to extend and retract. Typically, they are composed of a steel cylinder with a block fixed to both ends. The block on the rod end of the cylinder carries the rod seal, rod bearing, extension bumper, and the retract port. The block on the opposite end carries the retract bumper and the extend port. The bumpers

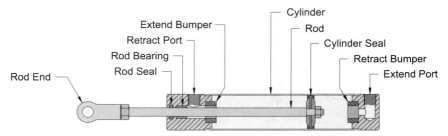

Figure 2-12 Double-acting cylinder.

are included to prevent the piston from slamming into the blocks during rapid extension and retraction. The rod bearing is usually in the form of a permanently lubricated, sintered bronze bushing. The rod seal can be either an O-ring or a sliding rod seal. The blocks can be affixed in two different manners. The first method, which produces a very reliable and compact design, is to swage the cylinder onto the cylindrical blocks. The second method (Fig. 2-13) is to retain the cylinder between two square blocks with four long bolts. Smaller cylinders are generally constructed using the swaging method whereas larger cylinders normally use the bolt-together method. The piston assembly usually carries a double-edge cylinder seal that is pinched between two load washers affixed to the rod via the rod nut.

A variation of the common cylinder is the spring return cylinder (Fig. 2-14). Rather than requiring air pressure to retract the cylinder, the unit carries a return spring. By simply venting the air pressure from the extend port, the cylinder retracts. A spring return cylinder is normally longer than a double-acting unit with the same stroke length. This is because the compressed spring must be accommodated during maximum extension. Pneumatic control systems can be considerably less complex when spring return cylinders are utilized.

A third variation is the double-ended cylinder (Fig. 2-15). This is little more than a standard cylinder with the rod protruding from both ends.

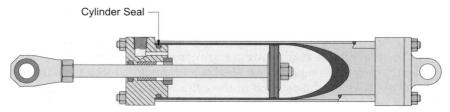

Figure 2-13 Bolt-together cylinder.

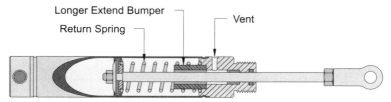

Figure 2-14 Spring return cylinder.

These units are not particularly common and most cylinder manufacturers only offer them for special requests.

There are only a few basic mounts that are usually supplied on pneumatic cylinders. Figure 2-16 shows a threaded nose mount that is common on smaller cylinders and a pivot mount, which is generally supplied on larger cylinders. Block mounts (Fig. 2-15) can be supplied as an integral part of the cylinder or as an accessory for the nose mount. Diaphragm and bolt-together cylinders (Figs. 2-13 and 2-17) are normally supplied with a universal bolt pattern that will accommodate a range of mount brackets and options. Most cylinders are supplied with a standard thread on the end of the rod. Any number of rod ends may be mounted to accommodate a wide variety of applications.

Controlling cylinders is not particularly difficult. The industry offers a wide variety of components that can be assembled to provide suitable control for almost any application. The simplest way to control a cylinder is to connect the air supply to the ports of the cylinder through a four-way ball valve. Turn the valve to one position and the pressure is vented from the retract side of the cylinder and applied to the extend side, extending the cylinder. Reverse the order and the cylinder retracts. Four-way valves are available in a wide range of designs, sizes, and actuator types. Among the actuator types are pedal, palm button solenoid, pneumatic, toggle, limit, push button, cable pull, and the like.

For most applications, it is desirable to control the speed at which the cylinder extends and retracts. Additionally, it is usually necessary to have different speeds for the extend and retract function. To provide

Figure 2-15 Double-ended cylinder.

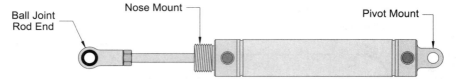

Figure 2-16   Nose and pivot mounts.

independent control, a needle valve, coupled with a check valve that feeds a bypass loop, is used. The needle valves control the amount of air that the cylinder receives, and in doing so the speed can be manipulated. The check valve and bypass loop allow the pressure to be freely vented when the cylinder is reversed. Figure 2-18 shows an electrically controlled cylinder installation with a four-way solenoid valve, needle valves, check valves, and bypass loops.

To simplify speed control applications, many manufacturers offer special valves, which incorporate a needle valve and check valve in a single fitting package. The speed control can range from a small slotted screw to a large comfortable knob. Figure 2-19 shows a manually controlled cylinder installation with a four-way palm valve and integral speed control valves.

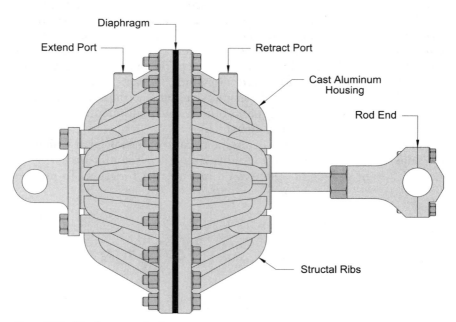

Figure 2-17   Diaphragm cylinder.

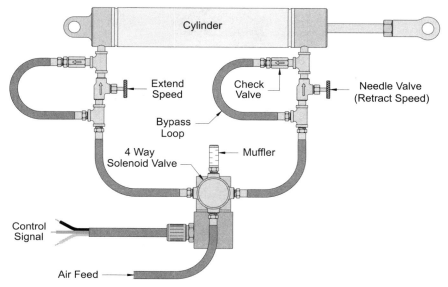

Figure 2-18 Electrical cylinder control.

When using a spring return cylinder, a three-way valve is utilized. A three-way valve is configured to connect the cylinder to either the air feed or the muffler by simple actuation. By depressing the pedal, air is directed to the extend port of the cylinder and the cylinder extends. Releasing the pedal will allow the air in the cylinder to vent through the muffler and it will retract. These valves are available in all the same sizes and with all the same actuator options as four-way

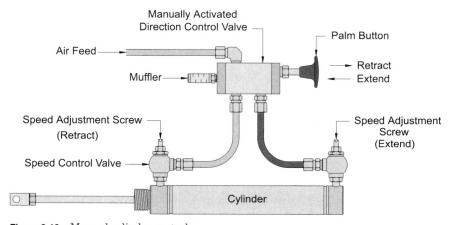

Figure 2-19 Manual cylinder control.

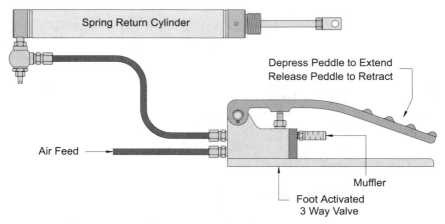

Figure 2-20   Spring return cylinder control.

valves. Figure 2-20 shows a foot-activated spring return cylinder installation with a three-way valve.

A wide range of force can be realized with pneumatic cylinders. The amount of force that a cylinder can produce is a function of the area of the piston times the air pressure. There are certain practical limits to consider when selecting a cylinder size. As an example: if a 4-in.-diameter cylinder is selected for producing a force of 100 lb, then a pressure of only 7.9 psi would be required. If the pressure regulation varies ±0.5 psi, then the force will vary 12.6 lb or about 12 percent. If a 1.25-in.-diameter cylinder is selected, then a pressure of 82 psi is required. A ±0.5 psi variation in the pressure will only cause a 1.2 lb variation in the force or about 1 percent. Similarly, if a 0.625-in.-diameter cylinder is selected, then a pressure of 326 psi would be required to generate a 100 lb force. This is a pressure that is much higher than most compression systems produce and higher than most pneumatic cylinders are rated for. Figure 2-21 provides a quick reference for matching a cylinder size, force generation, and delivery pressure for your application. One further consideration when using double-acting cylinders is the loss of piston area due to the rod diameter. During extension, the air pressures acts against the entire area of the piston. During retraction, the air acts against the area of the piston minus the area of the rod. For cylinders that are designed specifically for pneumatics, this loss is normally rather low because the rod diameter does not endure particularly high loads. However, if a hydraulic cylinder is placed into pneumatic service, then the rod loss can be significant. Hydraulic cylinders generate much higher loads than pneumatic cylinders of comparable size. Therefore, their rods are considerably larger in diameter than their pneumatic counterparts and must be considered during force calculations (Fig. 2.22).

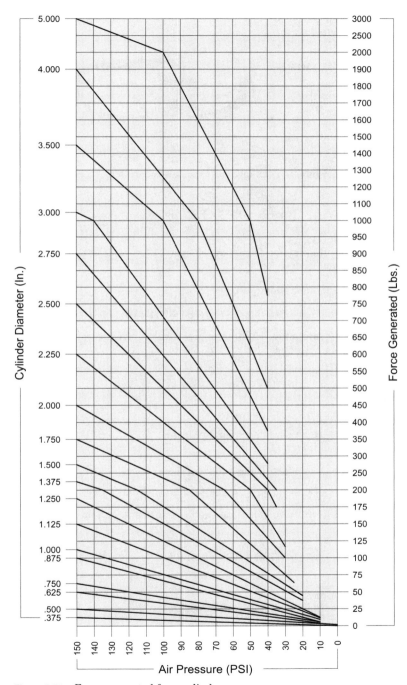

Figure 2-21   Force generated from cylinders.

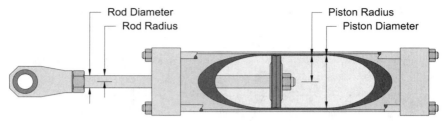

Figure 2-22 Piston diameter.

Use the following formula to calculate the precise force from a cylinder and pressure combination.

$$[\pi \times (RP \times RP)] \times PSI = FG$$

where π = 3.1416
    RP = radius of piston
    PSI = air pressure (in pounds per square inch)
    FG = force generated

Example: 1.75-in. cylinder and 85 psi

$$[3.1416 \times (0.875 \times 0.875)] \times 85 = 204.45 \text{ lb.}$$

Use the following formula to calculate the pressure required to generate a particular force from a cylinder.

$$[\pi \times (RP \times RP)] \div FR = PSI$$

where π = 3.1416
    RP = radius of piston
    PSI = air pressure (in pounds per square inch)
    FR = force required

Example: 2.25-in. cylinder producing 350 lb. of force

$$[3.1416 \times (1.125 \times 1.125)] \div 350 = 88 \text{ psi}$$

Use the following formula to calculate the force loss associated with the rod diameter.

$$[\pi \times (RR \times RR)] \times PSI = FL$$

where π = 3.1416
    RR = radius of rod
    PSI = air pressure (in pounds per square inch)
    FL = force lost

Example: 1-in. cylinder at 75 psi produces 57 lb of force on extension. Its rod is 0.25-in. diameter.

$$[3.1416 \times (0.125 \times 0.125)] \div 75 = 3.7 \text{ lb of lost force or about 5 percent}$$

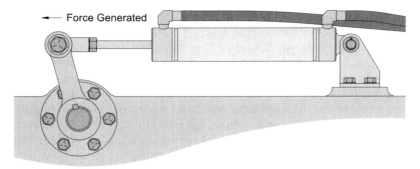

Figure 2-23   Cylinder installation.

Pressure Amplifiers

A variation of the simple cylinder is the pressure amplifier. These are also referred to as pressure intensifiers, pressure boosters, gas boosters, and hydraulic amplifiers. These units are designed to produce high secondary pressures from relatively low compressed air pressures. The most common of these devices generally consists of a large pneumatic cylinder pushing a smaller hydraulic cylinder (Fig. 2-23). These devices are generally a low-cost alternative to standard high-pressure pumps when high flow rates are not a consideration. Figure 2-24 shows a schematic representation of a pressure amplifier.

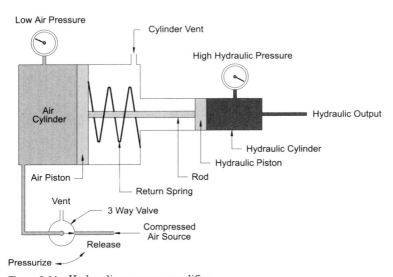

Figure 2-24   Hydraulic pressure amplifier.

The principle of operation is based on the force applied to the area of the larger air cylinder being transferred to the smaller area of the hydraulic cylinder, creating a higher pressure. The pressure is then a function of the differential piston areas. For example, if the air piston has an area ten times greater than the hydraulic piston, then applying 125 psi of compressed air will produce 1250 psi of hydraulic pressure.

Pressure amplifiers are used extensively for hydraulic clamping applications and are found in most machine, fabrication, and manufacturing facilities. Figure 2-25 shows a typical hydraulic clamping system utilizing a pressure amplifier. The clamping is controlled with the three-way valve at the upper left. The cylinder vent is equipped with a small filter to prevent piston fouling. The hydraulic cylinder is equipped with a

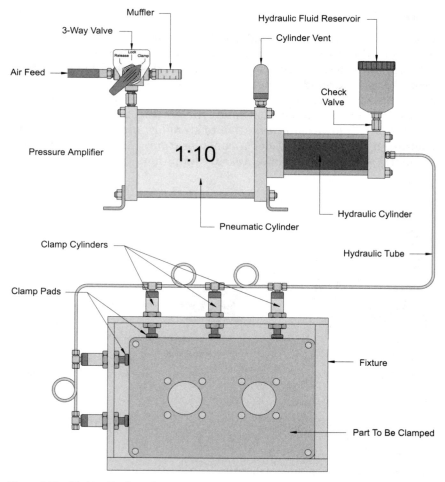

Figure 2-25 Hydraulic clamping system.

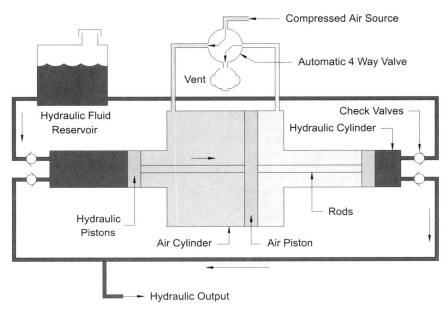

Figure 2-26 Pneumatic/hydraulic pump.

fluid reservoir to replace any losses during operation. If the pressure amplifier can be mounted directly to the fixture, then the hydraulic lines should be steel tubes, otherwise they must be hydraulic hoses. To prevent overdrawing the hydraulic cylinder, the clamp cylinders should be as small as possible. It should also be noted that the pressure ratio of many pressure amplifiers is boldly printed on the body of the unit as shown in Fig. 2-26.

Pressure amplifiers can also be set up to operate as hydraulic pumps for applications that do not require particularly high flow rates. In the case of a hydraulic pump, a second hydraulic cylinder is added, and the pneumatic cylinder is operated as a double-acting unit. An automatic four-way valve is used to control the cycle of the pump. Hydraulic fluid is supplied to the cylinders from a reservoir, and the flow is controlled via the check valves. Figure 2-26 shows a pressure amplifier set up as a hydraulic pump.

Percussion Tools

Another use of the cylinder is for percussion tools. These tools are little more than a cylinder that is set up to vent its air charge at maximum extension, then reset and repressurize. Common percussion tools include scalers, riveters, jack hammers, tampers, bolt cutters, chipping hammers, vibrators, chisels, and the like. Figure 2-27 shows a

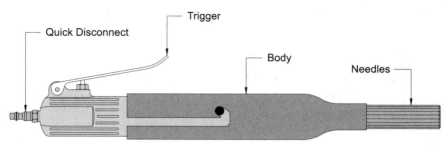

Figure 2-27 Needle scaler.

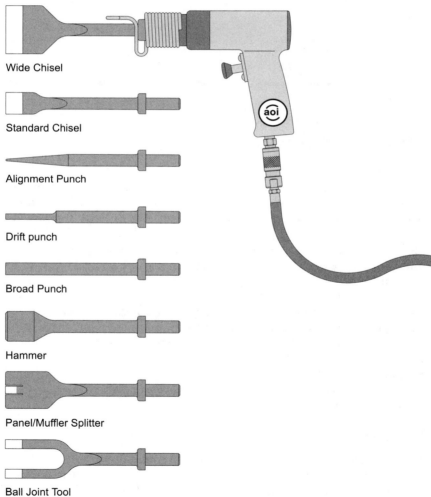

Wide Chisel

Standard Chisel

Alignment Punch

Drift punch

Broad Punch

Hammer

Panel/Muffler Splitter

Ball Joint Tool

Figure 2-28 Air hammer set.

needle scaler used for removing weld flux, spatter, rust, scale, and/or paint from metal surfaces. These units usually have a quick-release body that allows easy replacement of the needles. Figure 2-28 shows a typical air hammer and chisel set, which is appropriate for automotive use. These are commonly found in muffler and front-end shops where they are used to split exhaust pipes and dislodge ball joints. This type of air hammer is very inexpensive and will normally be found in any mechanic's toolbox. Figure 2-29 shows a typical jackhammer. You have probably seen jackhammers used for breaking up concrete by local road and utility crews. These tools are very powerful and are not for the faint hearted. Figure 2-30 shows a rivet/bolt cutter. The chisel point is placed at the base of the head of the rivet or bolt to be cut. When the tool is activated, the hammer action chisels off the head. The chisel is replaced with a flat punch, and the rivet or bolt is driven out of the workpiece. These tools are very common in the bridge, marine, and boiler main-tenance industries.

Air Motors

Air motors are almost always delivered in the form of a rotary vane device. These motors are similar to a rotary vane compressor (dis-cussed in Chap. 3), except that they use compressed air to generate rotation rather than using rotation to produce compressed air. The rotary vane motors are very simple and reliable devices. When used and maintained properly, they can provide a compact power source with an exceptional service life. They are utilized in almost every type of rotational pneumatic tool, from die grinders to overhead cranes.

Figure 2-31 shows a sectional illustration of a rotary vane air motor. The motor consists of a cylinder with an internal rotor. The rotor is placed off-center so that its outside diameter is tangent to the cylin-der's inside diameter. The rotor carries four vanes that are placed into slots. The vanes are usually spring-loaded and slide within the slots so that they are always in contact with the inside diameter of the cylinder. Compressed air is introduced into a cavity, which drives the rotor during expansion. As the rotor vanes rotate, the expanding volume progresses toward the vent. When the expanding volume reaches the vent, the air is allowed to vent. These motors normally operate at a rather high speed, and while some tools operate at the motor RPM, others incorporate some sort of gearbox to reduce the RPM and increase the torque to more suitable parameters for their application. Figure 2-32 shows a typical air motor. It should be noted

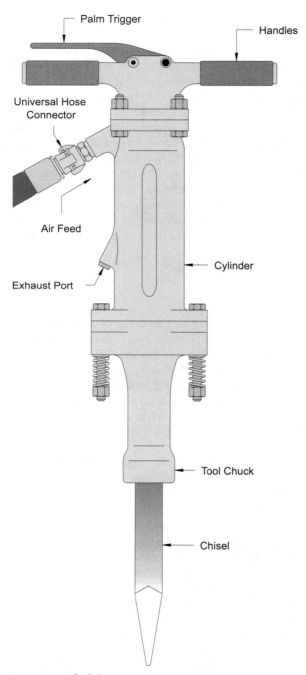

Figure 2-29   Jack hammer.

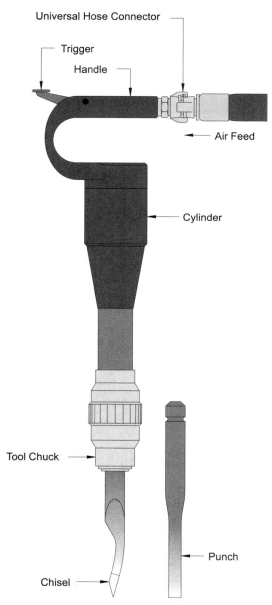

Universal Hose Connector

Trigger

Handle

Air Feed

Cylinder

Tool Chuck

Chisel

Punch

Figure 2-30 Rivet/bolt cutter.

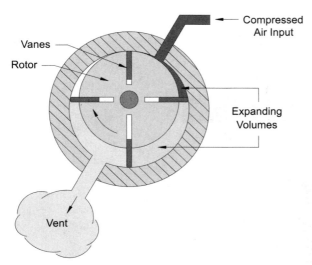

Figure 2-31   Rotary vane motor.

that most air motors are reversible by simply switching the locations of the air feed and muffler.

It should be noted that there are some piston motors being manufactured for special applications. These types of motors are generally not the preferred choice because of the high purchase price, maintenance requirements, and life expectancy.

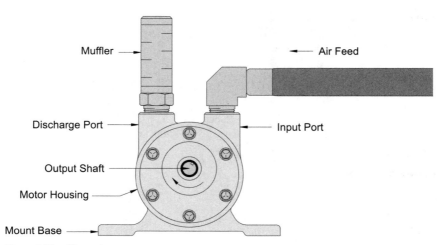

Figure 2-32   Air motor.

Rotating Tools

Another common use for compressed air is rotating tools. This category includes air wrenches, die grinders, disk sanders, mixers, polishers, belt sanders, drills, and the like. These types of tools are extremely common in the automotive and mechanical repair industries and can be found in virtually every auto repair shop in the United States. Rotating tools are almost always driven by a rotary vane air motor (see preceding section). Figure 2-33 shows a die grinder. These tools operate at a very high RPM and are quite handy for fitting and shaping parts. Figure 2-34 shows an air ratchet. The value of these tools is that they can rapidly screw and unscrew fasteners. Because air ratchets do not have internal hammers, it is necessary for the operator to do the initial loosening and final tightening of the fastener. Figure 2-35 shows a 1/4-in. drill motor. These units are very compact, lightweight, and operate at a high RPM. Figure 2-36 shows an angle sander. These units generally use a 7-in. sanding disk running at 5000 to 7000 RPM. Disk sanders are indispensable for blending sheet metal assemblies, weld joints, and castings. They are quite common in paint and body shops, where they are used to prepare surfaces and shape body filler.

Air motors do not produce particularly high torque. They normally derive their power through high RPM. Therefore, if an application has a high torque requirement then some sort of reduction must be used. Because of the high motor RPM, chains and belts are not preferred and a compact gear assembly is typically utilized. Figure 2-37 shows a representation of an RPM reduction drive. The motor drives a small primary gear, which, in turn, drives a larger secondary gear that carries the output shaft. The speed of the output shaft then provides an RPM and torque value, which is a function of the gear reduction ratio.

Figure 2-38 shows a heavy-duty drill motor. These tools generally use a planetary gear drive in a compact housing between the air motor

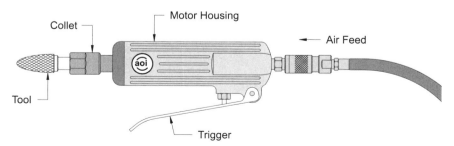

Figure 2-33 Die grinder.

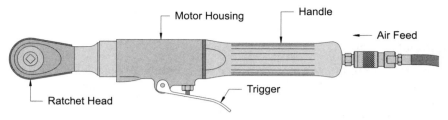

Figure 2-34 Air ratchet.

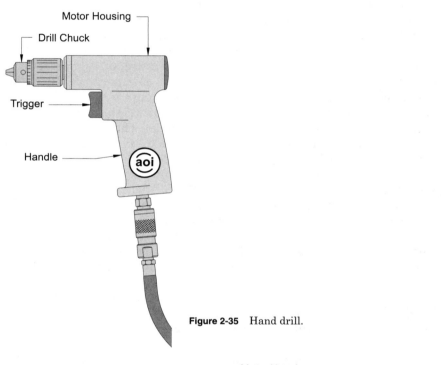

Figure 2-35 Hand drill.

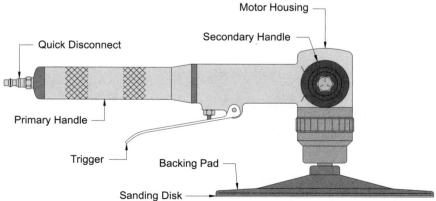

Figure 2-36 Disk sander.

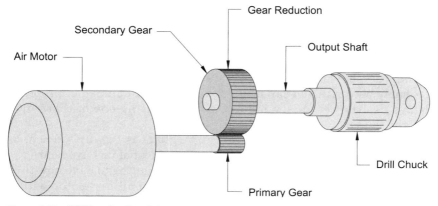

Figure 2-37 RPM reduction drive.

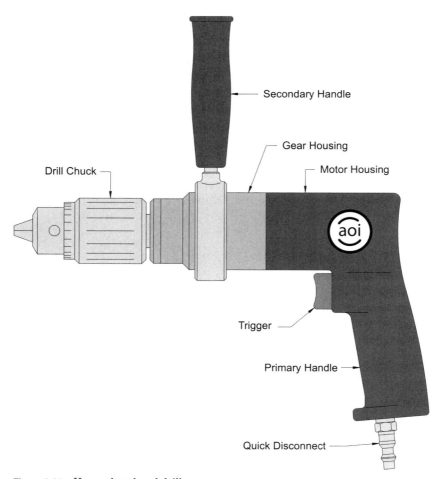

Figure 2-38 Heavy duty hand drill.

and chuck. Drill motors of this type are usually very compact and have a deceptive appearance. To the novice, they look like they might not be very powerful at all. Do not be deceived, these tools are extremely powerful and can be quite dangerous if handled improperly. Always use the secondary handle when operating one of these drill motors.

One interesting variation of the rotating tool is the air wrench, or impact wrench as they are sometimes called. Most of us have heard the repetitive hammering of these wrenches while waiting to have new tires put on our cars. An air wrench is usually a small air motor with a hammer mechanism built into the drive system. These tools provide two very important functions to the mechanic. The first is that they can produce extremely high torque values, and second they can operate at rather high RPMs. This allows the mechanic to loosen, tighten, screw, and unscrew fasteners at a very fast pace.

Figure 2-39 shows a stylized illustration of an air wrench drive system. The motor drives a cage, which carries two rollers. The rollers are allowed to float within slots cut into the cage. The rollers roughly engage the two anvil lugs on the output shaft. As the motor turns, the rollers force the output shaft to turn at the same RPM. The hammer action is provided when the rollers are allowed to shift into the secondary ID of the hammer. When the rollers reach the primary ID during the course of their rotation, they are forced to change their radius and, in doing so, they impart a high momentary force onto the anvils. This high momentary force is then transferred to the output shaft where it is utilized to tighten the fastener. Figure 2-40 shows a typical 1/2-in. -drive air wrench.

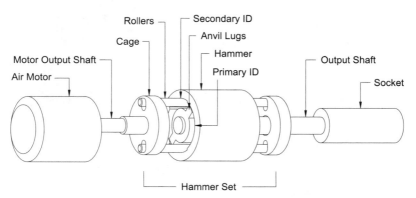

Figure 2-39   Impact wrench mechanism.

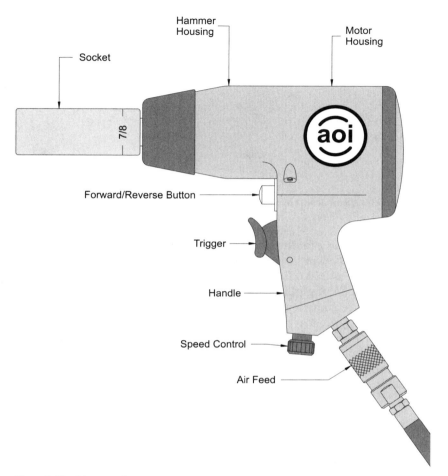

Figure 2-40 Air wrench.

Turbine Motors

For extremely high-speed applications, turbine motors are incorporated. These devices derive their power from operating at a very high RPM and, consequently, produce very low torque. These motors are ideal for applications that can utilize high RPM. They are extremely simple devices, which have only one moving part and two bearings. Dental drills or hand pieces as they are referred to are the most common use for turbine motors. The motor is a very small assembly at the tip of a handle. Figure 2-41 shows a typical dental hand piece. The feed line is connected to the back of the handle via a threaded connector. The feed line usually provides compressed air, water, and a fiber optic link. The water is sprayed onto

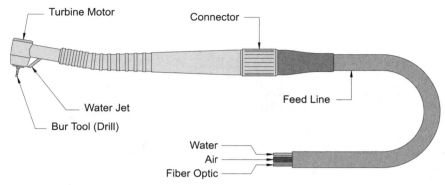

Figure 2-41   High speed dental drill (hand piece).

the tooth to cool the grinding operation. The fiber optic link is to provide light directly to the work area. A tool like the one shown may operate at a speed as high as 350,000 RPM! Other uses for turbine motors are in grinding and polishing glass, ceramics, industrial diamonds, and gemstones. There are some lower speed devices on the market that use turbine motors for their drive element. These devices must have a gearbox to lower the RPM and increase the torque of the motor. In general, conventional air tools that use turbine motors don't seem to provide any substantial improvement over their rotary vane counterparts and have the distinct disadvantage of being considerably more expensive to purchase.

Paint Sprayers

Spray painting is another common use for compressed air. Spray painting guns are manufactured for a broad range of coating application, from decorating tee shirts to painting cargo ships.

The simplest paint sprayer is the airbrush (Fig. 2-42). These guns are a type of siphon feed spray gun that is used for precise, fine work.

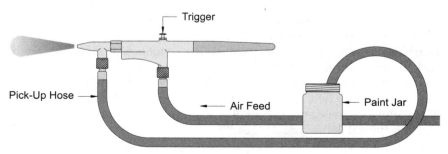

Figure 2-42   Air brush.

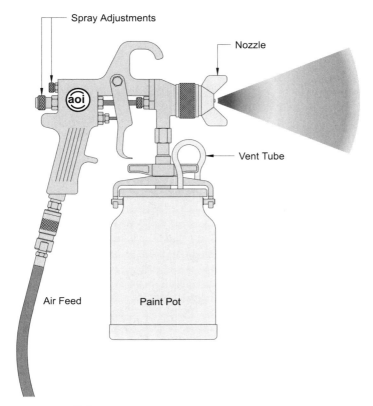

Spray Adjustments

Nozzle

Vent Tube

Air Feed   Paint Pot

Figure 2-43 Paint gun.

Airbrushes find favor with artists, auto detailers, tee shirt companies, photo retouchers, model builders, and the like. They are about the size of a fountain pen and, as such, they are very handy to use.

One of the most common types of spray gun is shown in Fig. 2-43. This particular unit has a 1-qt pot with siphon feed as an integral part of the gun. These guns are available as low-volume, high-pressure (conventional) or high-volume, low-pressure (HVLP) units. Conventional units are preferred for fine finishes, such as automobile painting. The HVLP units are designed to minimize overspray and are preferred for coating for general purposes, such as equipment and marine painting.

Another type of paint sprayer that is gaining popularity is the gravity feed gun. Figure 2-44 shows a typical gravity feed spray gun. Like other guns, these are available in conventional and HVLP units. The most noteworthy feature is the small paint reservoir on the top of the gun, otherwise these guns operate in the same fashion as other units.

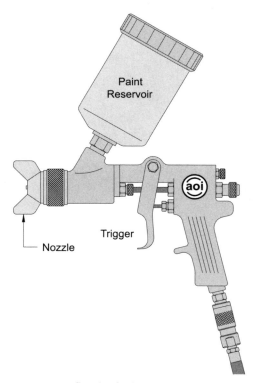

Figure 2-44   Gravity feed spray gun.

Most spray guns require a delivery pressure in the 15 to 30 psi range, so a line regulator is typically used in conjunction with the air feed.

For production painting, a pressure feed spray gun is desirable. These units are available in conventional or HVLP units and are typically delivered with an oversize tank for holding a large quantity of paint. Production paint tanks, shown in Fig. 2-45, are readily available in sizes ranging from 2 qt to 10 gal. The tank is filled with paint and its lid is clamped on. A pressure regulator is mounted on the top, which is used to adjust the paint delivery pressure. The tank is a pressure vessel and, as such, must be equipped with a safety relief valve that is set to open at a pressure no higher than the maximum rating of the tank.

Siphon Guns

A variation of both the blow-off gun and the paint gun is the siphon gun. These units are used for spraying low-viscosity fluids such as insecticides, degreasing agents, surface treatments, water/detergent mixes, and the like. Figure 2-46 shows a sectional view of a siphon gun. They

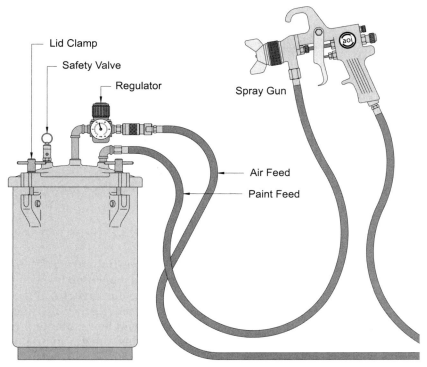

Figure 2-45 Production paint tank.

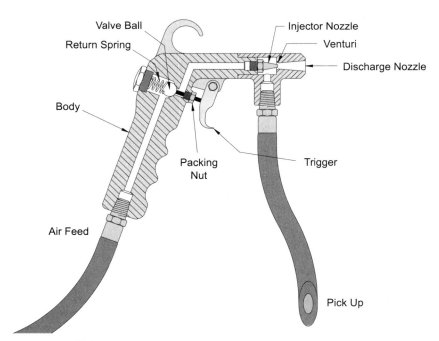

Figure 2-46 Siphon gun.

are little more than a blowgun with a Venturi added after the discharge nozzle. The Venturi creates a vacuum, which, in turn, draws up a fluid from the pickup tube. The pickup tube is typically placed into a bucket, which contains the fluid to be sprayed. The amount of fluid that is flowing can be adjusted by tuning the air delivery pressure.

These units can be very handy items to have in your toolbox. They are particularly popular with maintenance personnel who use them to quickly degrease equipment they must repair. Caution should be exercised when using siphon guns with small compressors. These units require a substantial amount of airflow and care should be taken not to overwork the compressor.

Sandblasters

Sandblasters are used extensively in the coating industries. These machines are available in sizes ranging from small, benchtop blast cabinets for intricate parts, to large units intended for structural and marine blasting. Sandblasting will effectively remove any substance that may be adhering to the surface of the material being blasted. In addition to sand, there are a variety of other blast media available including glass beads, walnut shells, steel shot, abrasives, and the like.

Figure 2-47 shows a schematic representation of a typical pressure-feed sandblasting unit. Sand is held in the hopper, which is pressurized. The sand flow control valve is used to adjust the quantity of sand that is introduced into the air stream. The sand is carried, in suspension, through the hose and is discharged out of the nozzle.

Sandblast cabinets are a convenient way to bring blasting into the shop. These machines are typically self-contained units with recycling blast media. Figure 2-48 shows a floor standing blast cabinet. The cabinet is a glove box with air, media, and exhaust feed throughs. There is typically a florescent light fixture on the top of the cabinet that illuminates the inside. The blast media is placed into a cone shaped reservoir, which is the lower section of the cabinet. For easy media removal and replacement, the reservoir is generally equipped with a spring loaded dump door at the bottom. These units typically use a siphon gun, which draws the blast media up from the bottom of the reservoir. The dust that is invariably generated is drawn into the dust collector through the exhaust port and hose. The compressed air feed is controlled by the operator via a floor pedal valve.

The part to be blasted is placed on the deck grate inside the cabinet. The door is closed, and the operator slides his hands into the protective gloves and picks up the blast gun. The blast is turned on by simply pressing the floor pedal. Blast cabinets of this nature

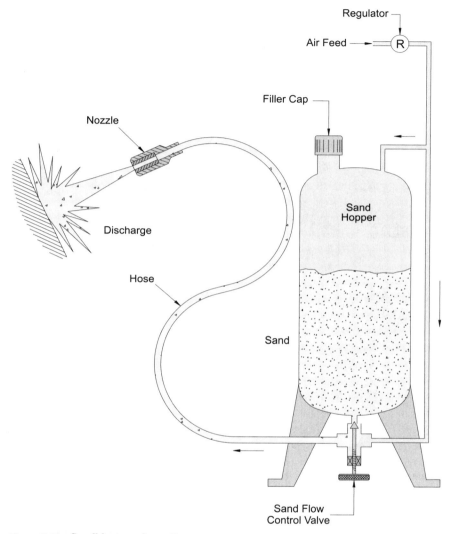

Figure 2-47 Sandblaster schematic.

usually work best with medium grit glass beads. Glass beads will produce a particularly fine finish and are generally preferred for intricate parts.

Tumble blasters are generally used for the production blasting of parts that are otherwise too small to handle. The blasting takes place in a rotating barrel. A load of small parts is placed into the basket,

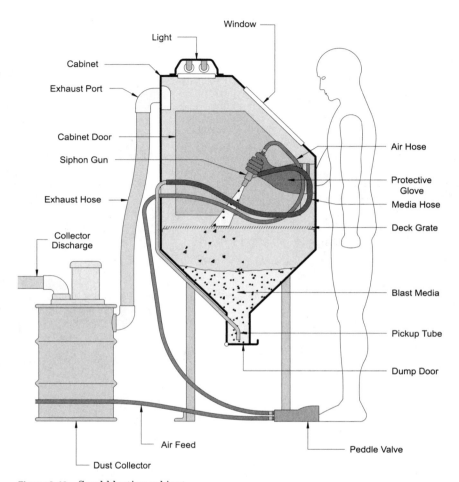

Figure 2-48 Sand-blasting cabinet.

which is inserted into the barrel. A lid is affixed to the top of the barrel, which has a siphon blast gun mounted in a bearing to allow rotation. The barrel is rotated at about 30 RPM by a drive motor and belt. As the barrel rotates, the parts inside are tumbled and blasted. The blast media pass through the basket and back down into the media reservoir. These units generally use the same media handling system as a standard blast cabinet. Figure 2-49 shows a stylized schematic of a typical tumble blaster. These units are particularly useful in blasting small components, such as electrical connector inserts or clock parts.

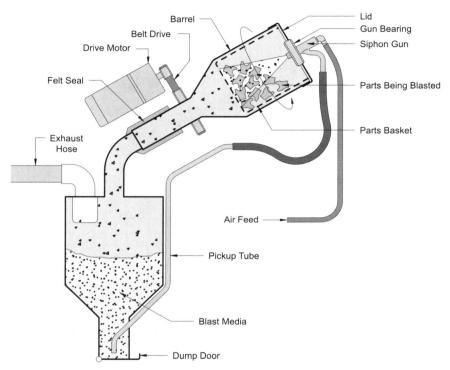

Figure 2-49   Tumble blaster.

Air Nailers

The woodworking and home construction industries are now dominated by air nailers, a tool that was virtually nonexistent 30 years ago. These tools have become an integral part of any business that deals with wood. This is apparent when one is passing by a new home construction site. Where once the sound of hammer blows were heard, now the hissing snap of air nailers ring out.

These tools come in a wide variety of configurations that are suitable for all sorts of applications. Air nailers are available for framing, finishing, roofing, general carpentry, brad nailing, flooring hammers, and the like.

Air nailers are essentially air cylinders controlled by a pulse valve. When the trigger is pulled, a piston is driven down via compressed air. It picks up a nail from the magazine, forces it into the workpiece and resets. In addition to the standard trigger, air nailers usually have a "nose" or "bump" trigger. The nose incorporates a safety tip that will not

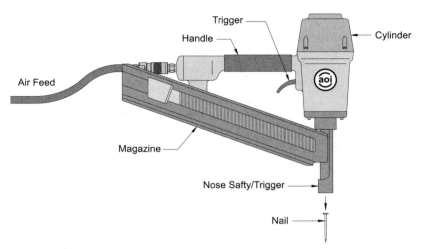

Figure 2-50   Air nailer.

let the gun fire unless the nose is in contact with the workpiece. If the trigger is depressed and held, the nose safety then becomes a nose trigger. The gun can then be bumped into the workpiece and a nail is fired each time. This provides for very fast nailing. An experienced carpenter can drive 10 to 20 nails in the same amount of time it used to take him to hammer just one. Figure 2-50 shows a typical framing nailer, which you might find on any home construction site or at your local home improvement center.

Air Staplers

Pneumatic staplers are found in a great number of applications. They are just as commonplace as air nailers, but are used in slightly different industries. The standard pattern stapler (Fig. 2-51) is the cornerstone tool of the upholstery industry. Upholsterers typically use three different size staples, which are referred to as finish, standard, and large. A typical upholstery shop will have a number of these units distributed throughout the facility. The furniture industry makes extensive use of pneumatic staplers for basic fastening and for the temporary clamping of glue joints. Cabinetmakers use pneumatic staplers as a matter of course. The staple provides a joint that is just as strong as a nail but with a much smaller fastener. There are a number of roofing guns on the market which utilize stapling technology rather than nails. When applying cardboard panels to wooden shipping crates, staples are the preferred method.

A standard pneumatic stapler is principally the same device as an air nailer. The unit will have a cylinder which is controlled by a pulse trigger.

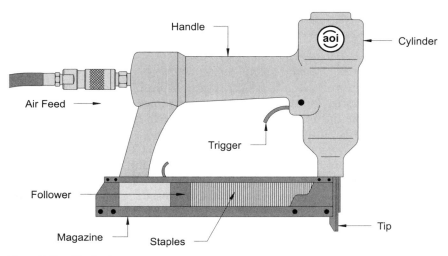

Figure 2-51   Air stapler.

The cylinder is connected to a hammer which picks up a staple and drives it into the workpiece. Releasing the trigger will reset the unit. Unlike the air nailer, staplers rarely incorporate nose or bump triggers.

There are a vast number of specialty staplers offered on the market. Units are constructed for everything from binding to construction. Although these specialty units are always referred to as staplers, they may not share the same general appearance as a standard unit. As an example, Fig. 2-52 shows a box stapler. This unit is designed to close a packed box with just a couple of trigger pulls. When the trigger is pulled,

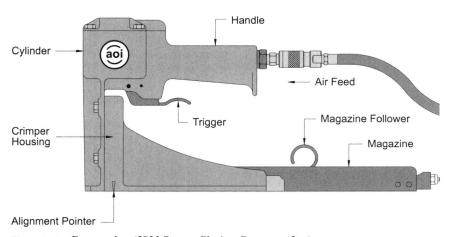

Figure 2-52   Box stapler. (*ISM Carton Closing Company, Inc.*)

the staple is inserted into the box lid and a set of jaws fold or crimp the staple on the far side of the cardboard.

Caulk Guns

For production work, a manual caulk gun is simply inadequate, which makes pneumatic caulk guns the preferred choice. These guns are very basic, reliable, and inexpensive. They are little more than a pressure chamber with a charge of caulk and a trigger style needle valve. Figure 2-53 shows a typical pneumatic caulk gun. The pressure housing is unscrewed from the handle assembly and a tube of caulk is dropped in. The pressure housing is screwed back onto the handle assembly and the tip end of the cartridge is forced into a pressure seal. To operate the gun, the trigger is depressed allowing pressure to be applied to the inside of the housing, which, in turn, pushes the caulk out of the cartridge tip. To control the gun, a variable valve design is used to allow the operator to adjust the caulk flow rate using the relative position of the trigger. Trigger valves also incorporate a venting function, which relieves the housing pressure when it is released.

Grease Guns

Most of us have dealt with a grease gun at one time or another. Manual units are just fine for occasional machine and automobile service work. However, in situations where there are a number of grease points that must be serviced on a day-to-day basis, a pneumatic grease gun provides

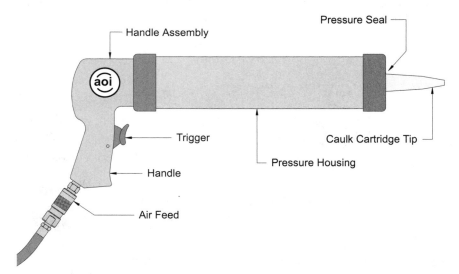

Figure 2-53   Pneumatic caulk gun.

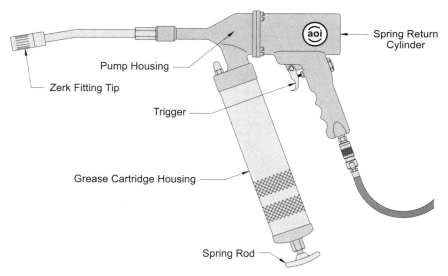

Spring Return
Cylinder

Pump Housing

Zerk Fitting Tip

Trigger

Grease Cartridge Housing

Spring Rod

Figure 2-54   Pneumatic grease gun.

a great advantage. These units do not differ much from a manual gun except that compressed air provides the pump power. A standard grease cartridge is placed into the cartridge housing in much the same way that it would be placed into a manual unit. Each time the trigger is depressed the cylinder extends and cycles the grease pump within the pump housing. When the trigger is released, the cylinder and pump reset. Caution should be exercised when using air powered grease guns. If the gun is triggered when the tip is not on a zerk fitting, a stream of grease will shoot out across the shop. This will really tick off any coworker who might accidentally be in the path of the flying grease. Figure 2-54 shows a typical pneumatic grease gun.

Pop Rivet Puller

Pneumatic pop riveters are exceptionally nice tools to have. They streamline sheet metal assemblies to a degree that only someone that has experienced one of these units can appreciate. Although there are manual pop rivet guns, they are very slow to operate and can only handle smaller sizes of rivets. As a matter of practicality, pop rivets are intended to be used with power riveting tools. Pneumatic pop riveters are simple, reliable, and inexpensive tools. When you use one for the first time, you'll wonder why you haven't had one all along. The units are very user friendly. Simply insert the mandrel of the rivet into the tip of the riveter, insert the rivet into the workpiece and pull the trigger. The

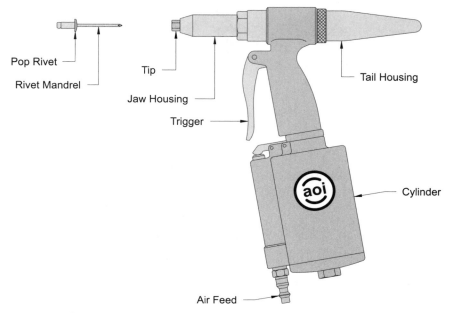

Figure 2-55   Pneumatic pop riveter.

jaws within the jaw housing grab the rivet mandrel, pull the rivet and snap the mandrel off. Releasing the trigger resets the tool for the next operation. The separated mandrels are collected in the tail housing and must be periodically emptied. Figure 2-55 shows a common pop riveter.

Dual Diaphragm Pumps

Pneumatic diaphragm pumps are a variation of the diaphragm cylinder. These pumps are commonly used on construction sites and in field applications where electricity is not readily available. They are also popular in the food processing industries because of their general washdown characteristics. It is common to find these types of pumps with maintenance personnel at petrochemical plants because of their durability and inherent safety. Since they have no electrical components, there is no fire or ignition hazard associated with the pumps. Their durable construction allows them to pump foodstuffs, oils, grease, paint, muddy water, sludge, slurries, trash, rotting vegetation, and the like. They are commonly found at sewer and street repair sites and can be heard buzzing away alongside the construction compressor.

The pump consists of two diaphragms that are connected with a connecting rod. Compressed air is introduced into the pneumatic cavity,

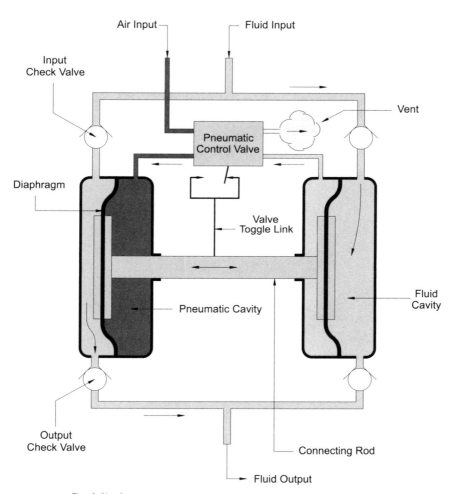

Figure 2-56   Duel diaphragm pump.

which pushes the diaphragm away. This, in turn, pulls the opposite diaphragm, which draws fluid into its fluid cavity. At the peak of the stroke, a toggle valve vents the air from the pneumatic cavity and redirects the compressed air to the opposite cavity. This reverses the motion of the diaphragms and the fluid is forced through the pump's output. Figure 2-56 shows a schematic representation of a dual diaphragm pump.

Vacuum Generators

Compressed air may be used to provide a vacuum by incorporating a vacuum generator. These units are simple Venturi pumps that are driven

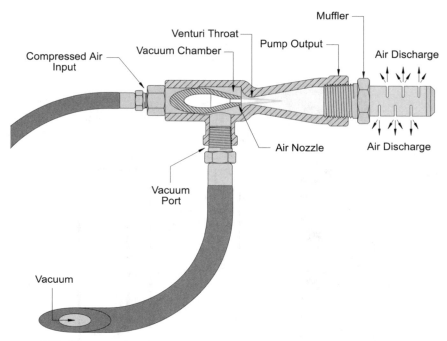

Figure 2-57   Vacuum generator.

with compressed air. Venturi vacuum pumps are not a very efficient way to generate a vacuum; however, they have a number of advantages over more traditional pumps. First and foremost, they have no moving parts and are exceptionally reliable. In addition, they are compact, have no electrical requirements, are very inexpensive, and require no maintenance.

Figure 2-57 shows a sectional view of a typical vacuum generator. To lower the sound level, the pump output of the vacuum generator is generally equipped with a muffler. If fluids or hazardous gases are to be pumped, then the muffler can be replaced with a connection to an appropriate dump site.

Air Bearings

Compressed air can be utilized to create extremely low-friction, high-load bearings. An air bearing is simply an inversely placed cup that compressed air is directed into. The lifting capacity of the bearing is directly related to the area of the bearing times the compressed air pressure. These devices can be very powerful tools. As an example: a 12-in.

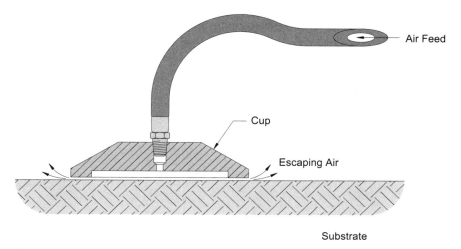

Figure 2-58 Air bearing.

-diameter bearing that has a pressure feed of just 90 psi can lift over 10,000 lb! Four of these units can lift a 20-ton piece of equipment with almost zero friction.

One of the drawbacks is that the substrate on which the bearing operates must be very smooth, level, and free of any blemishes. Additionally, the clearance between the bearing face and the substrate is rather small. On the plus side, the air escaping from the bearing has a tendency to blow away any dirt that may be in its path.

Air bearings can be made in a flat version, as shown in Fig. 2-58, or they can be incorporated in rotational applications. Shaft air bearings have the advantage of being self-cooling and are oftentimes used for extremely high RPM applications.

Spray Misters

Spray misters are rather common devices in most machine shops. They are a particularly effective way to apply coolant to metal cutting operations. Spray misters are popular because they do not require the costly support equipment that a flood system needs. Additionally, they generally don't make as big a mess as a flood system does.

These devices are little more than a Venturi with air and coolant control valves. They blast out a fine mist of coolant, which is very effective in penetrating the cutting area. Figure 2-59 shows a typical spray mister for machine operations.

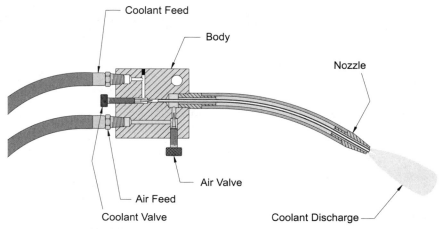

Figure 2-59   Spray mister.

Vibrators

Pneumatic vibrators are normally supplied in two different versions. Figure 2-60 shows a rotary ball vibrator. These units have a circular track that a ball can freely traverse. Compressed air is introduced into the housing and the ball is forced to rotate around the tract. The frequency of the vibration is controlled by varying the air pressure. The amplitude of the vibrations is controlled by the weight of the ball. Figure 2-61 shows a piston vibrator. These devices operate in essentially the same manner as an air hammer. The air pressure can be adjusted to control the frequency of the device. The amplitude is controlled by either changing the weight of the piston or by damping the motion of the piston. Piston vibrators can have either hard or soft stops. Hard stops increase the amplitude of the device and provide a

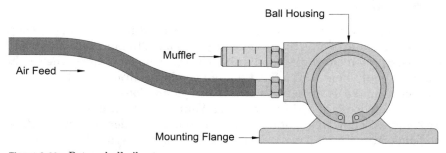

Figure 2-60   Rotary ball vibrator.

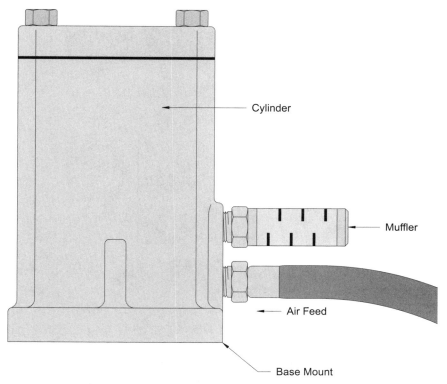

Cylinder

Muffler

Air Feed

Base Mount

Figure 2-61   Piston vibrator.

sharper vibration. Soft stops decrease the amplitude and the vibrations are less sharp.

Automobile Lifts

Nearly every auto repair shop in the country has automobile lifts in their service bays. These lifts vary in mechanical design, but their power source is generally an air-over-hydraulic system. These units operate on the same principle as the air-hydraulic amplifiers (see earlier section on Pressure Amplifiers).

Figure 2-62 shows a schematic representation of an in-ground auto lift. A large hydraulic cylinder is embedded into the ground. The end of the column is fitted with a set of adjustable lift points that adapt to a wide variety of cars. The cylinder is plumbed to an air cavity through a restrictor orifice. As air pressure is applied to the air cavity, the cylinder rises and, in turn, lifts the car. When the air is released from the cavity, the cylinder lowers. The restrictor orifice is a safety device, which limits the speed at which the cylinder can be raised and lowered.

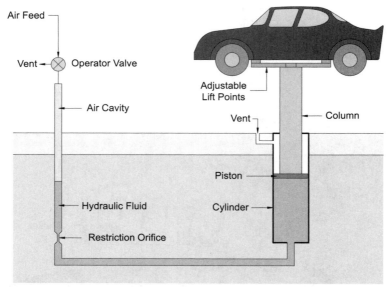

Figure 2-62 Automobile lift.

Air-Over-Hydraulic Jacks

Most of us are familiar with hydraulic jacks. The basic bottle jack is used for all sorts of applications, from automobile service to foundation repair. Most hydraulic jack designs are available with a pneumatic pump in addition to the manual pump. The pneumatic pump is simply an air-hydraulic pressure amplifier that is fitted as an integral part of the standard jack.

Figure 2-63 shows a schematic diagram of an air-over-hydraulic bottle jack. The jack functions in the same manner as a fully manual unit, except it also has an air port and control valve. The pressure amplifier is placed in parallel to the manual pump. In this manner the amplifier and the manual pumps can utilize the same check and lower valves. The amplifier is equipped with a return spring and a three-way palm valve. To operate the jack, the palm valve is depressed and the amplifier pumps a charge of hydraulic fluid into the main cylinder that is equal to a single stroke of the manual pump. When the palm valve is released, the air pressure is vented and the amplifier resets. By pressing and releasing repeatedly, the palm valve jack can be raised to any height that the mechanic desires. If a higher level of sensitivity is required, the mechanic can use the manual pump.

Figure 2-64 shows a commercial air-over-hydraulic jack, which is available from many auto repair supply houses. In addition to the pneumatic function, these units will generally retain all the normal

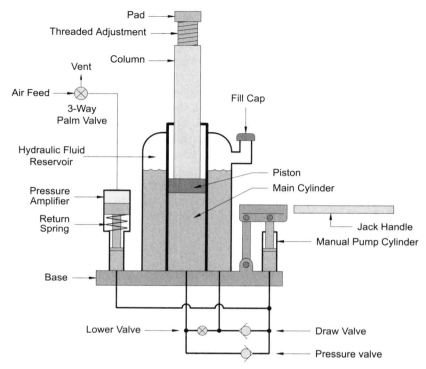

Figure 2-63   Air-over-hydraulic bottle jack schematic.

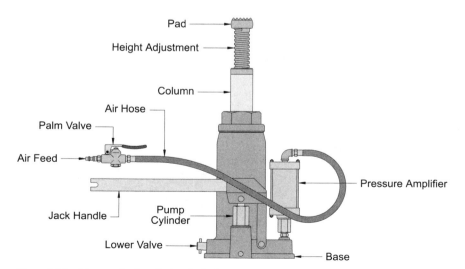

Figure 2-64   Air-over-hydraulic bottle jack.

features of a standard hydraulic jack. Note that the palm valve is equipped with an industrial quick disconnect, which allows easy connection to the air system. Lowering the jack is accomplished in the same manner as a standard jack, by opening the lower valve with the end of the jack handle.

Barrel jacks are essentially the same system as a bottle jack. These jacks are generally of higher capacity than their bottle jack counterparts. Barrel jacks are generally used in large truck, construction equipment, or railroad service shops. Many of these units are pneumatic operation only while some retain a manual pump. The wheel set is usually spring loaded to raise the jack for easy movement. When the jack starts to lift, the base settles onto the ground to provide an extremely stable support.

Figure 2-65 shows a typical barrel jack with pneumatic-only control. The lower control is in the form of a button, which allows precise control of the jack.

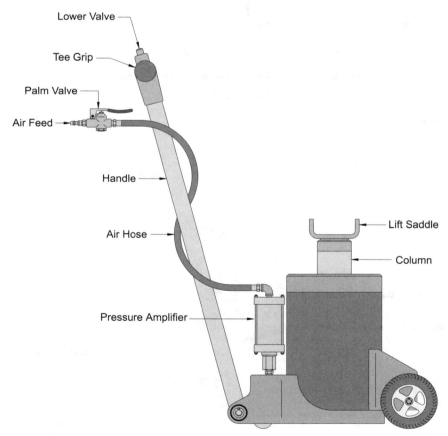

Figure 2-65   Barrel jack.

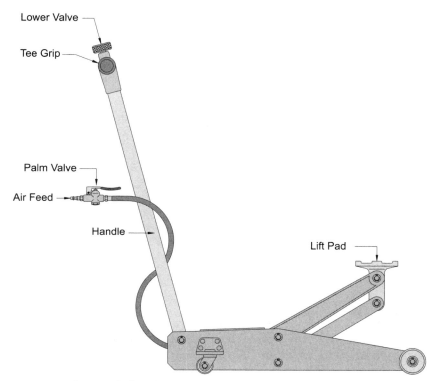

Figure 2-66 Pneumatic floor jack.

Floor jacks are very common in the automotive service industry. These jacks are an essential part of any well-equipped shop. Adding a pneumatic function to these jacks just improves their versatility. Pneumatic versions work in the same manner as the bottle jack and retain all the standard features of a manual floor jack.

Figure 2-66 shows a typical air-over-hydraulic floor jack. To the casual observer there is little difference in the appearance of the pneumatic and standard versions.

Air Bag Jacks

Air bag jacks generally consist of a bag or bladder that is inflated with compressed air to produce lift. The cross section of the bag allows these jacks to produce very high lift loads with relatively low pressures. There are some units sold that will lift an automobile with just the pressure that the engine exhaust produces. In addition to high lifting capacities, these jacks are generally very light weight and are oftentimes

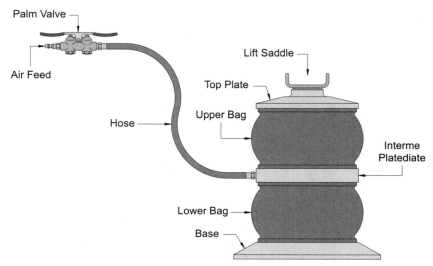

Figure 2-67   Bag jack.

utilized in situations where equipment weight is a consideration. Because the jack is made of rubber bladders, the input pressure is generally limited to less than 100 psi. To assure that the bag(s) do not rupture, most of these units incorporate some sort of pressure relief valve.

Figure 2-67 shows a typical air bag jack. This particular unit uses two bags interconnected with an intermediate plate. The lift saddle is mounted into the top plate within a swiveling bearing. The air input is through the intermediate plate and is controlled with a three position three-way palm valve. The jack is shown in the raised position.

Pneumatic Suspension

The most common use for air bags is as a suspension component. Many heavy trucks use auxiliary air bags in the role of load leveling. As an example, a 6-yard dump truck may be able to carry a 20-ton load of dirt. The ride characteristics of the truck when it is empty versus when it is fully loaded will be dramatically different. Auxiliary air bags can be installed to compensate for this range. When the truck is empty, the bags are deflated and the leaf springs properly support the vehicle. When the truck is loaded, the air bags are inflated and add considerable load-carrying capacity to the chassis. In this manner, large trucks can have the best of both worlds.

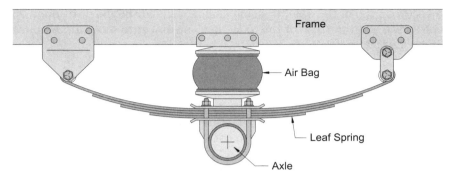

Figure 2-68 Leaf spring suspension with auxiliary air bag.

Figure 2-68 shows a leaf spring suspension system that incorporates an air bag for auxiliary load compensation. The air bags are generally installed between the top of the axle and the frame. Many trucks come standard with this feature, those that do not can be retrofitted with relative ease.

Another popular use for air bags is what is commonly referred to as air ride suspension. In this situation, the entire vehicle is supported on air bags. This arrangement is common on cross country buses and passenger trains. An air bag suspension will normally provide a smoother ride than metal springs. This is because the air bags will dampen high frequency noise and vibration, which will pass through a metal spring. Another attribute of an air ride suspension is that the vehicle can be lowered for passenger loading and unloading and raised into a run position when the bus is underway.

Figure 2-69 shows a typical air ride suspension system. In this case, a swing arm locates and controls the axle and the air bag is placed

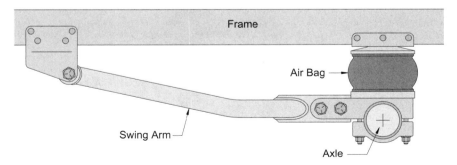

Figure 2-69 Air ride suspension.

directly on top of the axle. In some designs, the air bag is positioned on the swing arm, in front of the axle, so it is subject to a certain amount of mechanical advantage.

Air Brakes

Any vehicle in the United States with a gross overall weight of 20,000 lb or higher is required by the Department of Transportation to have a redundant braking system. For most trucks, this is in the form of air brakes. The most important attribute of an air braking system is that the brakes fail on. That is to say, if the brake control system fails, the brakes are automatically applied. This is an important safety feature on any truck that has a high gross vehicle weight.

Figure 2-70 shows a typical air brake arrangement. The axle carries a drum, which is equipped with two internal shoes. The bottom of the shoes is located on the shoe pivot and the top is located with the cam. The two shoes are held in place with two keeper springs. The cam is connected to the cam arm on a common shaft. A spring loaded diaphragm cylinder is used to rotate the cam via the arm and shaft. In a no pressure situation, the cam rotates so that the shoes are pushed apart and come in contact with the drum. To release the brake, air pressure is applied to the cylinder, the cam rotates back into a neutral position, and the shoes release the drum.

Figure 2-71 illustrates a basic air brake system. The system is fed by an engine-mounted pump, which feeds a main receiver. The main receiver feeds two different brake circuits, the front wheels and the

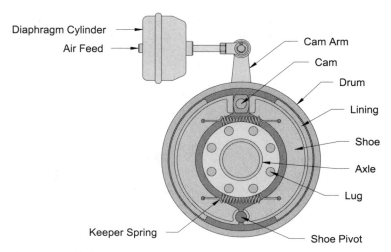

Figure 2-70 Air brake.

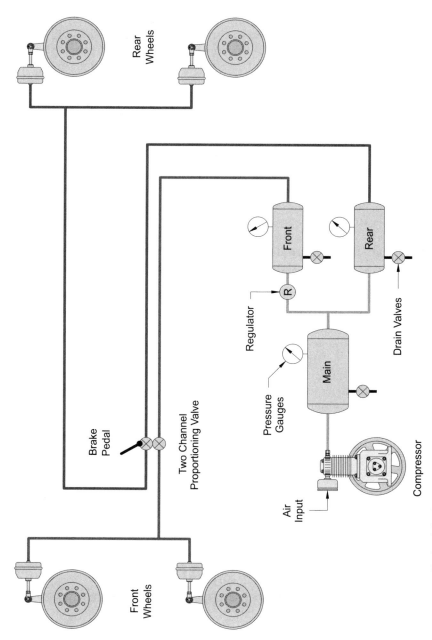

Figure 2-71 Basic air brake system.

back wheels. Because the front wheels do not have the same load as the rear wheels, a regulator is used to reduce the pressure on the front wheel circuit. Both circuits are controlled through a two channel proportioning valve, which is actuated through the brake pedal. When the pedal is depressed, the air in both circuits is released in proportion to the position of the valve. When the pedal is released, full system pressure is returned to the brake cylinders.

In the mid 1800s George Westinghouse invented what is arguably the most significant technological advancement in railroad history—the air brake. He received his first air brake patent in 1869 and the railroad industry was changed forever.

Air brakes are a fundamental standard on modern trains. Before the advent of air brakes, a brakeman would have to jump from car to car and apply the hand brake wheels to control the speed of the train! Needless to say this was a very dangerous profession.

Modern train air brakes are actually simpler in design than a truck's system. However, like trucks, the number one attribute of a train air brake system is that it applies the brakes when it fails.

Each car carries a single air cylinder, which actuates the brake shoes through a series of levers and connecting rods. There is one brake shoe, which is applied directly to the outside diameter of each wheel. Figure 2-72 shows a standard brake arrangement on a typical railroad car.

Figure 2-73 shows a schematic diagram of a train air brake system. The system can be divided into two sections—the air source on the locomotive and the car system. An engine-driven compressor supplies compressed air to a main receiver. The output of the air is controlled by the brake valve, which feeds the brake pipe. At either end of the brake pipe is an angle cock.

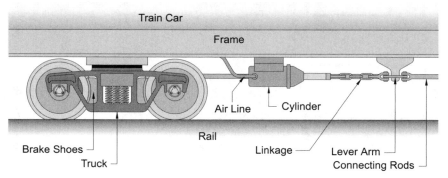

Figure 2-72 Train car air brakes.

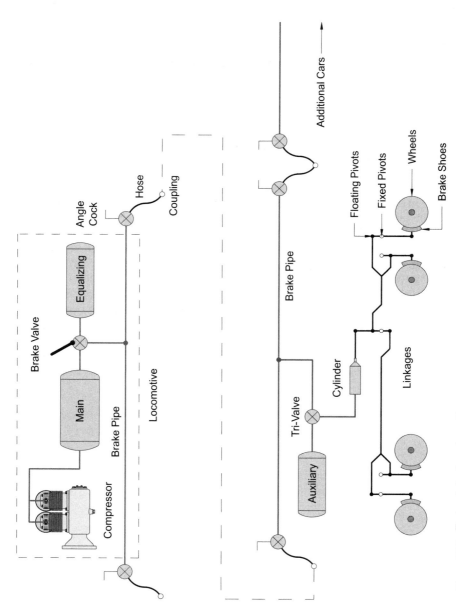

Figure 2-73 Train air brake schematic.

Every rail car made carries a brake pipe, tri-valve, auxiliary receiver, and brake cylinder. As with the locomotive, the brake pipe has an angle cock on each end. When the engineer applies the brakes, the pressure is released from the brake pipe and the cylinder allows the shoes to come in contact with the wheel. When the brakes are released, the pressure in the brake pipe is allowed to return to full system pressure and the cylinder retracts the brake shoes. The couplings between cars are designed to disconnect if the car becomes uncoupled. When this happens the system pressure is vented, the brakes are fully applied, and the train comes to a complete stop.

Self-Contained Underwater Breathing Apparatus (SCUBA)

Self-contained underwater breathing apparatus (SCUBA) packs are more of a compressed air system than they are a single application. The basic function of this arrangement is to provide breathing air to the diver while underwater. However, the pack must also provide at least three additional functions: variable delivery pressure based on dive depth, buoyancy compensation, and backup or emergency air.

Figure 2-74 shows a schematic representation of a typical SCUBA arrangement. The tank is charged with a special high-pressure compression system, which is discussed further in Chap. 12. The tank is usually a 3000-psi aluminum cylinder with a 80-ft^3 capacity, and has a special valve and an O-ring port. A pressure gauge reads the tank pressure and, therefore, provides the diver with a relative reference for his or her downtime. The output of the tank is generally fitted with a

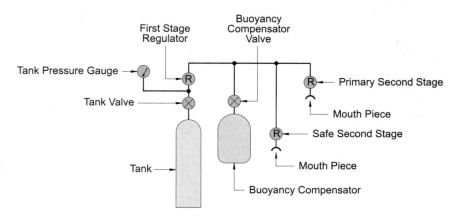

Figure 2-74   SCUBA system schematic.

pressure compensated first-stage regulator. A pressure compensated regulator will increase its output pressure as the diver descends and reduce its output pressure during ascent. The output of the first-stage regulator feeds at least three different devices. First, and most important, is the diver's second-stage regulator. Second-stage regulators are extremely sensitive to pressure compensation. They must be sensitive enough to deliver air only when the diver breathes in and stop feeding air during exhaling. The second function is the safe second-stage regulator. This regulator provides redundancy for the primary second-stage regulator and can provide emergency air to a fellow diver in the event of a catastrophic equipment failure. The third function is to inflate the buoyancy compensator. Because the diver's buoyancy can vary at different depths, a bladder is filled with enough air to counter the diver's tendency to sink.

Figure 2-75 shows a typical SCUBA pack. These systems are generally set up as an integral unit, which is worn by the diver as a backpack.

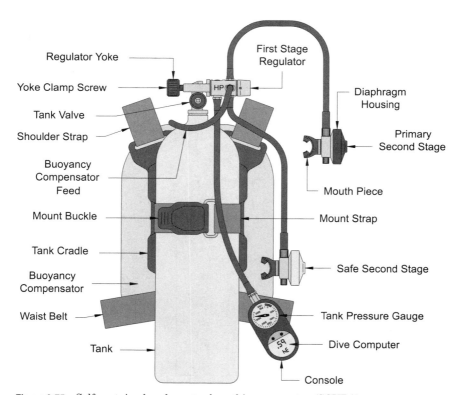

Figure 2-75   Self-contained underwater breathing apparatus (SCUBA).

The pressure gauge is usually mounted on the end of a hose, which allows the diver to easily read the tank pressure. The output of the first stage is generally set up with several ports arranged around the body. This setup is usually referred to as an octopus.

Self-Contained Breathing Apparatus

Much like SCUBA packs, self-contained breathing apparatus (SCBA) packs are more of a compressed air system than they are a single application. The basic function of a SUBA pack is to provide breathing air to personnel in environments that do not have suitable air quality. The most visible application of SCBA is with fire departments. Oftentimes, while watching the evening news, SCBA packs can be seen on the backs of firemen.

Figure 2-76 shows a schematic representation of a typical SCBA system. The tank is charged with the same type of high-pressure compression system used to charge SCUBA tanks. Modern tanks are usually thin wall steel cylinders that are reinforced with a carbon fiber wrap. Cylinders are manufactured in this manner to limit the weight of the pack. The cylinder valve is a two-part unit that has a globe type primary valve and a ball type quick acting valve. The latter is used on site and the former is used to preserve the charge of the tank during extended storage periods. The output of the tank is routed through an on-demand valve, which supplies air to the operator only when needed.

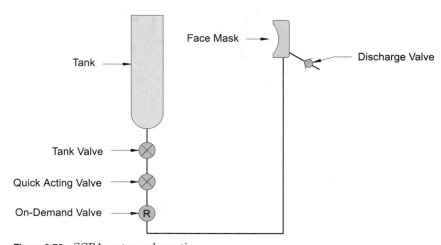

Figure 2-76   SCBA system schematic.

The mouthpiece is typically a full face mask which provides eye protection as well as breathing air.

Figure 2-77 shows a typical SCBA pack and a face mask. Notice that the face mask is equipped with a 6 strap head harness. This is done to assure that the mask doesn't get torn off during difficult situations.

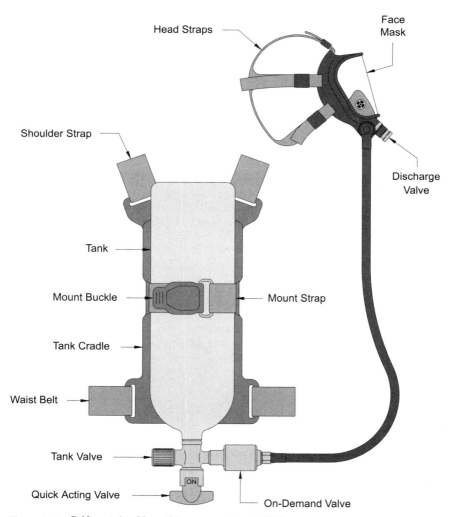

Figure 2-77   Self-contained breathing apparatus (SCBA).

Questions*

1. What is the most common pneumatic tool?

   (A) Air wrench   (B) Nail gun   (C) Paint sprayer   (D) Blowgun

2. What kind of motion does a cylinder generate?

   (A) Vibrating   (B) Linear   (C) Pumping   (D) Circular

3. What are the two principal cylinders?

   (A) Piston   (B) Vane   (C) Diaphragm   (D) Shaft

4. What type of motor is normally used in rotary tools?

   (A) Turbo   (B) Rotary vane   (C) Piston   (D) Screw

5. What common medias are used in sandblasters?

   (A) Sand   (B) Glass beads   (C) Steel shot   (D) Walnut shells

6. What air tool dominates the home construction industry?

   (A) Wrenches   (B) Air jets   (C) Vacuum generators   (D) Nailers

7. What type of pump is usually powered by compressed air?

   (A) Centrifugal   (B) Diaphragm   (C) Piston   (D) Metering

8. What is the foremost reason for which the vacuum generators are used?

   (A) Maintenance   (B) Expense   (C) Reliability
   (D) No moving parts

9. What are the two reasons for which air bearings are used?

   (A) Low friction   (B) High load capability   (C) Low cost
   (D) Clearance

10. Name two different types of vibrators.

    (A) Battery   (B) Piston   (C) Dust   (D) Rotary ball

_____

*Circle all that apply.

Compressor Types

Air compressors are designed and constructed to fulfill a broad range of applications, from filling swimming pool floats to feeding multimillion square foot manufacturing facilities. For most of us, our first real encounter with air compression was when we used a bicycle tire pump for the first time. We connected the pump to our low tire, pumped it up, and never stopped to consider that the tire pump is really a basic air compressor. If we look closely at the design, we can see that it is simply a cylinder that has been modified to operate as a single-cylinder compressor. As with all compressors, an outside power source is required. In the case of the tire pump that power source is you.

The bicycle tire pump (Fig. 3-1) is a very simple mechanism and, as such, is ideal to illustrate the principal components of an air compressor. The pump is made up of a steel tube fitted with a piston and rod. The top of the rod has a tee handle for interfacing with the power source (you). The bottom of the cylinder is equipped with a base that provides an output for the compressed air. When the rod and piston are forced down, the air trapped in the lower part of the cylinder is compressed. The compressed air is pushed through the discharge ports, the check valve, output hose, and then out of the air chuck. The check valve is necessary to stop the air in the tire from rushing back into the cylinder when the piston is lifted back up. As the piston is pulled up, a partial vacuum is formed in the lower part of the cylinder, air is drawn in around the leather seal, and a new charge of air is introduced into the compressor. This seal acts as both the piston seal and as an input check valve.

When I was growing up, we used tire pumps for a number of compressed air applications. My favorite use was an old army surplus tire pump that my brother and I modified with a long hose and an air nozzle. One of us would operate the pump while the other used the air nozzle

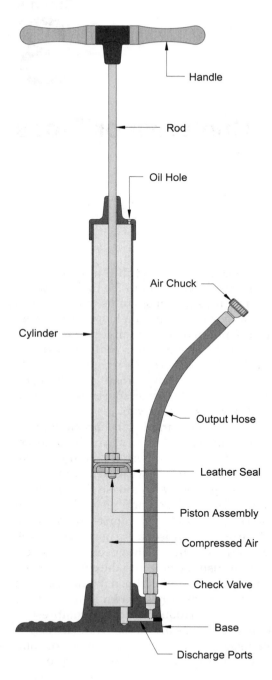

Figure 3-1 Bicycle tire pump.

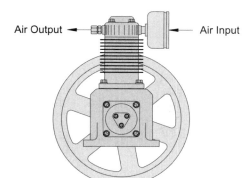

Air Output ← Air Input

Figure 3-2 Single cylinder piston pump.

to blow off dust and dirt from whatever project we were working on. Another popular use for the tire pump was recharging our water-filled fire extinguisher, or flumer as they are sometimes called. We would spend 30 minutes pumping up the fire extinguisher to a suitable pressure and then expend its contents in about 2 minutes.

Among the various compressor designs, there are two basic categories, the first of which are blowers. Although blowers are actually low-pressure compressors, they are not generally thought of as compressors. Blowers are usually reserved for moving air for ventilation applications. These devices may be utilized for applications ranging from pushing air through a heating, ventilation, and air conditioning (HVAC) system to providing air flow for the aeration of a fluid. They are oftentimes used to create a gas-based slurry to transfer grain stocks or powered materials. It is important to have a clear understanding of blowers, their uses, and limitations. In many industrial applications, compressed air is used when a blower would be a much better solution. Knowledge of this category can arm you with the necessary information to identify situations when compressed air may not be the best solution. The first part of this chapter provides a brief discussion on some of the more common blower types and a few of their applications.

The second basic category is high-pressure compressors. These units are generally positive displacement pumps, which can provide pressures in the 35- to 200-psi range. There are only six basic approaches to compressing air for general purpose compressed air applications. The second part of this chapter reviews each one of these different approaches in enough detail to provide you with a solid understanding of the various pump types.

Squirrel Cage Blowers

The first type of blower that we will discuss is the squirrel cage blower. These blowers are the most common blower type in the world. They are

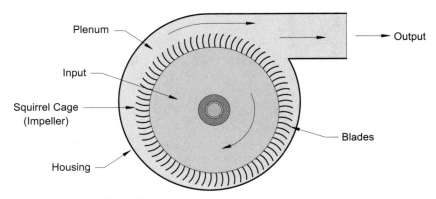

Figure 3-3 Squirrel cage blower.

found in the heating and ventilation system of nearly every home, manufactured in the past 60 years. These devices cannot produce any pressure to speak of, but make excellent prime air movers.

The blower incorporates an impeller, or squirrel cage, mounted
directly to the output shaft of an AC motor. The motor is imbedded into
the center of the impeller and carries special mounts that allow the
cage/motor assembly to be mounted directly to the housing. This allows
for a very simple and inexpensive assembly that may be readily mass-
produced. The impeller is a series of blades assembled in a drum pattern. As the cage is rotated, usually at 3400 RPM, the blades force air
from the input of the impeller to the plenum of the housing. The air in
the plenum is then forced through the output. Figure 3-3 shows a
schematic representation of a typical squirrel cage blower.

Centrifugal Blowers

Centrifugal blowers are commonly found in commercial heating, air-
conditioning, and ventilation systems. In this role they serve the same
function as a squirrel cage blower, as a prime air mover. A centrifugal
blower can produce a higher output pressure than a squirrel cage blower
and, therefore, is the preferred choice for pumping large positive pressure air distribution systems. These blowers are similar in construction
and operation to the squirrel cage units. The impeller is generally a flat
plate with four to ten blades arranged in a spoke pattern on one side.
The plate has a hub, which mounts the impeller to a shaft. Smaller
units will have their impellers mounted directly to the motor shaft.
Larger units will carry a drive shaft, which is driven via a v-belt drive.

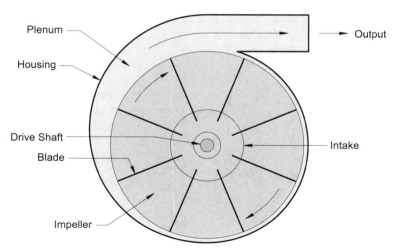

Figure 3-4   Centrifugal blower.

The housing is normally a welded assembly constructed from heavy gauge sheet metal.

As the impeller rotates, air is drawn into the input and slung outward toward the Outside Diameter (OD) of the impeller through centrifugal action. The air is forced into the plenum and through the output. Figure 3-4 shows a schematic representation of a typical centrifugal blower.

Regenerative Blowers

Regenerative blowers are used when a higher pressure is required. These types of blowers are oftentimes placed into aeration service because they can produce pressures as high as 3.5 psi. These types of blowers can also be used as a vacuum pump to provide a negative pressure for various commercial operations. Figure 3-5 shows a cutaway of an ordinary regenerative blower. The impeller is a heavy disk with a series of blades affixed to the OD in the form of spokes. The impeller is generally mounted directly to the motor shaft and the housing is mounted directly to the motor frame. As the impeller rotates, air is drawn in through the input. The air is forced out into the recirculation chamber through centrifugal action (Fig. 3-6). The air is forced back down the sides of the chamber and back into the roots of the blades. As this recirculation occurs, the air slips backward into the next blade cavity, adding to the pressure of that cavity. This recirculation or regeneration is continuously repeated until the air charge reaches the output and is allowed to escape.

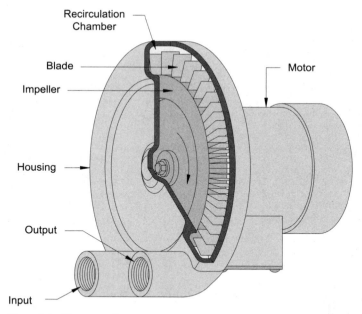

Figure 3-5 Regenerative compressor.

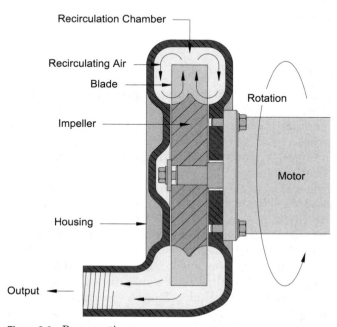

Figure 3-6 Regenerative compressor.

Rotary Lobe Blowers (Roots)

Rotary lobe blowers, or roots blowers as they are sometimes referred to, are a type of positive displacement pump that is used in many different applications. Like the regenerative blower, the roots blower is capable of producing several pounds per square inch of output pressure and, therefore, has certain applications where the design is indispensable. These types of blowers were originally designed to ventilate large mine shafts and can still be seen in this role. They were also used to provide positive pressure for industrial two-stroke diesel engines. Many of us have probably noticed the superchargers that are used on top eliminator dragsters; these are roots type blowers. Another application is providing roughing and first-stage pumping to support large industrial vacuum systems. These types of blowers are even used for gas flow measurement.

The blower consists of a pair of meshing rotors imbedded into a housing (Fig. 3-7). As the rotation occurs, the lobes mesh with one another and form a trapped charge of air. The charge of air is pushed around the inside of the housing and forced through the output. The two rotors are generally timed with a set of timing gears.

Liquid Ring Compressors

Liquid ring compressors generally fall into the "blower" category. They are a type of positive displacement pump which operate on the same

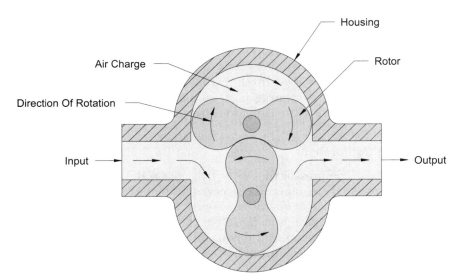

Figure 3-7 Two lobe (roots) compressor.

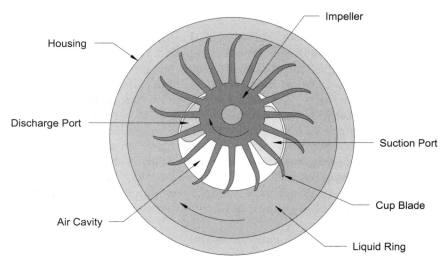

Figure 3-8 Liquid ring compressor.

principle as the rotary vane pump discussed later in this chapter. These pumps have an impeller, which is placed off center in a housing. The impeller has a series of cupped blades arranged radially around a hub. The housing has a charge of low viscosity liquid inside, usually water. As the impeller is rotated, the water is forced out to the housing inside diameter (ID) through centrifugal action and forms an air cavity at the center. The interaction of the impeller and the liquid ring forms a series of expanding and contracting cavities. As the cavity volume increases, it draws in air through the suction port. The impeller rotates the cavity to the other side and its volume starts to decrease. As the cavity volume decreases, the air is forced through the discharge port. Some liquid escapes through the discharge port, so a constant source of makeup water must be added to the ring at all times during operation.

Diaphragm Compressor

The first type of compressor that we will discuss is the diaphragm pump. These compressors are generally reserved for home and hobby use. The pump is rather simple in operation. Figure 3-9 shows a schematic representation of a diaphragm compressor. As the crankshaft rotates, it moves the diaphragm up and down, much like a piston. When the diaphragm is pulled down, air is pulled in through the input and into the compression chamber. As the diaphragm is forced up, the charge of air is compressed. As the diaphragm reaches the top of its stroke, it

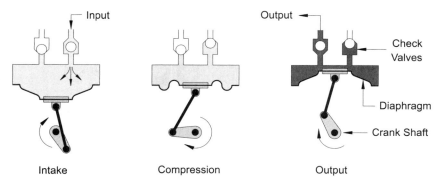

Figure 3-9 Diaphragm compressor cycle.

pushes the compressed air through the output. Output pressures of diaphragm compressors usually do not exceed about 60 psi. This is because their diaphragms cannot withstand high differential pressures or elevated temperatures.

Figure 3-10 shows a typical diaphragm compressor. The motor and crankcase are an integral assembly. The diaphragm and valve plate are normally bolted to the crankcase. Most diaphragm pumps can be reversed so that they may be used as a vacuum pump or a compressor.

Piston Compressor

The second type of system that we will examine is the piston compressor. This type of compressor is very similar to the tire pump. These are the most common small compressors used in a wide variety of applications, from home and hobby to small industrial plants. They are simple,

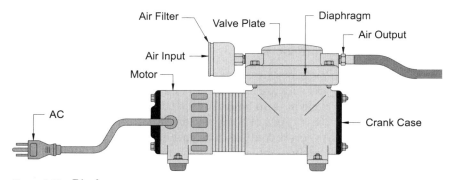

Figure 3-10 Diaphragm compressor.

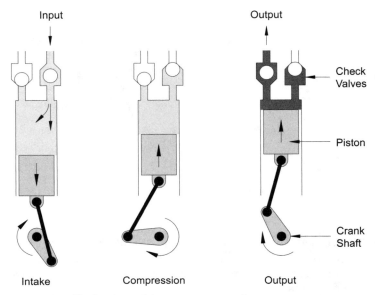

Figure 3-11 Single-stage piston compressor cycle.

inexpensive, and durable. They may be configured as small tank-less systems for lesser duties or they can be multi-horsepower packaged units as seen in auto repair facilities. Piston compressors function in much the same way as a diaphragm compressor, except they incorporate a piston with a cylinder as their compression element. Figure 3-11 shows a schematic representation of a piston compressor.

Figure 3-12 shows a typical single-cylinder compressor. This type of compressor is commonly found in the home garage shop or supplying air to the job site. Compressors of this type normally have a 10- to 20-gal receiver, output regulator, and wheel set configured as a complete package.

Double-Acting Compressors

Double-acting compressors are a variation of the simple piston compressor. These compressors are similar to a double acting cylinder in that their compression cycle acts on both sides of the piston. During the first half of the 1900s, these types of compressors dominated the large compression industry. As screw compressors gained popularity, large piston compressors were slowly displaced from the market. It is uncommon to find this type of equipment in operation today. However, there are a few of these units still in operation, dotted throughout industry and, if the

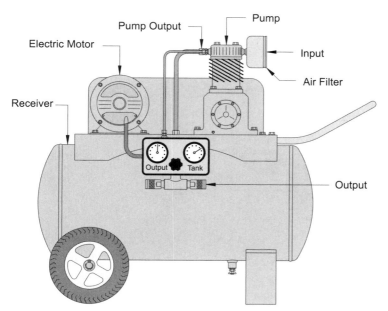

Figure 3-12   Single cylinder compressor.

opportunity presents itself, they are quite interesting to examine. For the most part, double acting compressors are relegated to specialized compression applications.

Figure 3-13 shows a schematic representation of a double acting compressor. The piston is a cylinder that is fitted with upper and lower heads. There are input and output check valves on both heads. When the piston travels up, the air charge is compressed and forced out through the check valve. At the same time, air is drawn into the chamber on the bottom of the cylinder. As the piston travel reverses, the cycle is flipped. The piston/cylinder/head assembly is generally mounted to a spacer section. This section provides facilities to mount two rod seals. The upper seal isolates the lower portion of the cylinder from the outside and the lower seal isolates the crankcase from the outside.

This arrangement makes double acting compressors "oil less." Larger units from the first half of the 1900s were typically lubricated, by spraying a mist of oil into the intake. The crankcase has a crankshaft, connecting rod and crosshead, which are usually pressure lubricated. It should also be noted that this method of isolating the cylinder assembly is utilized in manufacturing ultraclean, oil-less compressors.

Figure 3-14 shows a typical commercial single stage double acting compressor and motor mounted on a standard base. The crankcase and

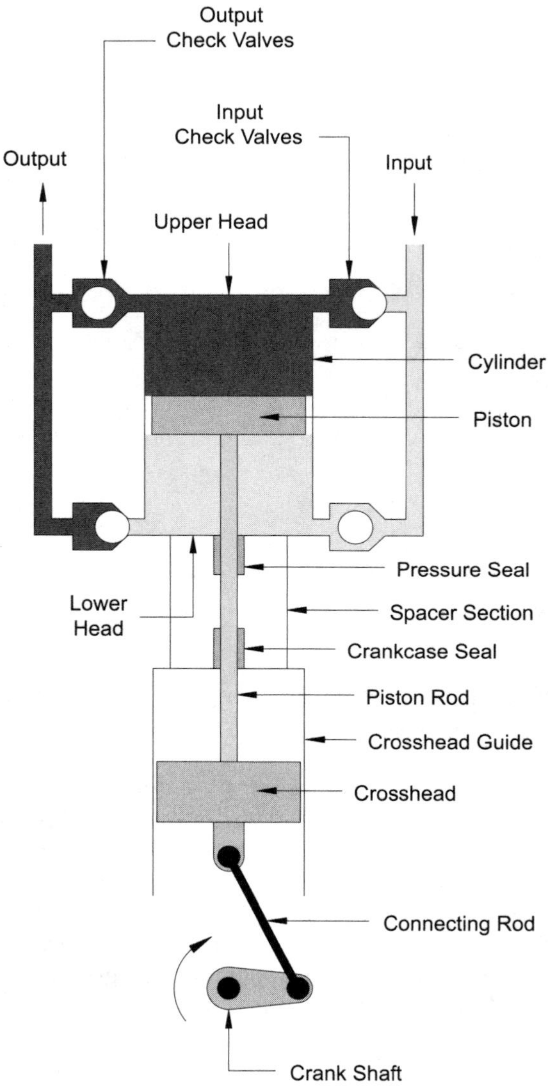

Figure 3-13   Double-acting compressor schematic.

crosshead guide are typically a standard piston pump that has been modified by replacing the piston with a crosshead. The crosshead guide is usually the stock cylinder block, which can be fitted with a specially manufactured spacer and cylinder assembly. These units are available in single or two-stage patterns.

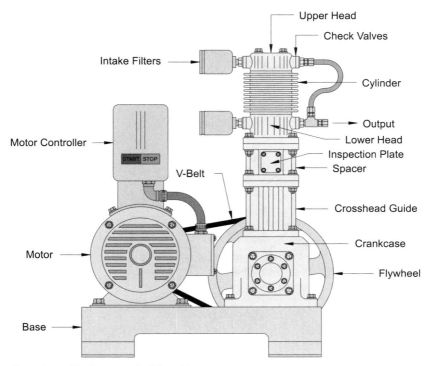

Figure 3-14 Single-stage double-acting compressor.

Two-Stage Compressors

The third type of compressor is a variation of the piston compressor—the two-stage compressor. A two-stage compressor compresses air in two separate steps. This provides a more efficient use of the input power and will typically produce lower-cost compressed air. These compressors are normally found in the 2- to 30-hp range and can be found providing at least some service in almost every industrial plant and service facility in the world. Figure 3-15 shows schematic representations of a two-stage compression cycle.

As the first-stage piston moves down in its bore, it draws in outside air. When the piston rises, it compresses the air charge. As the piston approaches the top of its stroke, the compressed air is forced out through the first-stage output and is fed into the intercooler. The charge is then drawn into the second stage and compressed a second time to the final output pressure.

The efficiency of a two-stage compressor comes primarily from the intercooler. The intercooler is used to remove the bulk of the heat that

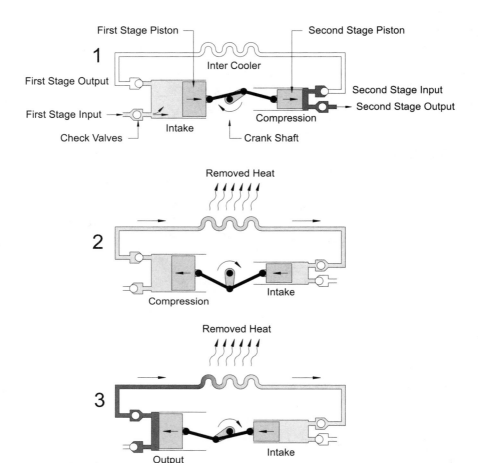

Figure 3-15   Two-stage piston compressor cycle.

is generated during the first-stage compression cycle. By removing the heat of compression, the air charge is condensed and, therefore, will provide more total air to the second stage. The more heat the intercooler removes, the more efficient the compressor will be.

Another attribute of two-stage compressors is that they generally output higher pressures than single-stage units. Normally, a two-stage compressor will produce 175 psi as its peak pressure. This is too high to feed into most general compressed air systems. To bring the pressure down to a usable level (about 90 psi), a master regulator is normally utilized. The big advantage of this arrangement is that the compressor will not cycle as much as a system that operates at a lower output pressure. The reason for this lower cycle interval is that the air in a receiver at

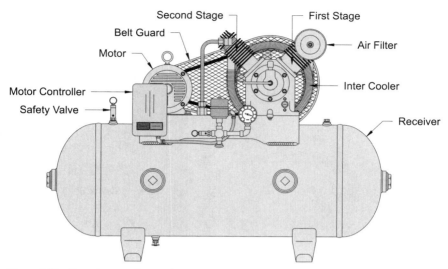

Figure 3-16   Two-stage reciprocating compressor.

175 psi has nearly twice the density, and therefore nearly twice the capacity, as the same receiver at 90 psi.

Figure 3-16 shows a typical two-stage packaged reciprocating compressor. These systems are commonly delivered in the 5- to 30-hp range and will normally be set up to shut off at 175 psi. For small businesses and applications that have a broad air delivery requirement, such as auto repair shops, these compressors are an ideal solution.

Figure 3-17 shows a typical two-stage pump that would be used on packaged compressors in the 5- to 15-hp range. Take note that the first-stage cylinder has a larger diameter than the second-stage cylinder.

Screw Compressor

The fourth compressor type to review is the screw compressor. These compressors dominate the compression industry in the 40- to 500-hp range. They are very common in almost every aspect of industry. They have the advantage of having a rather low purchase price and operating cost per standard cubic foot of air (SCFM). Their disadvantage is that they are complex, less reliable, difficult to repair, and have a lower life expectancy than their piston counterparts. Their cost is so low compared to a piston unit, that it is unusual to encounter a piston compressor over 30 hp and you may never encounter a piston compressor over 100 hp. In the 40- to 500-hp range, the screw compressor simply delivers more bang for the buck. Typically, screw compressors are single stage and produce output pressures in the 100- to 150-psi range.

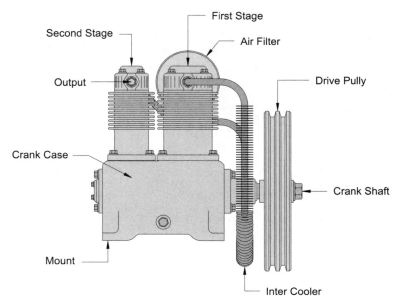

Figure 3-17   Two-stage pump.

The screw compressor is a rather simple mechanism in theory; in real life it gets a little more, messy. Figure 3-18 shows a stylized view of a screw compressor. These units are referred to as "screw" compressors because they have two different screws, which are contained within

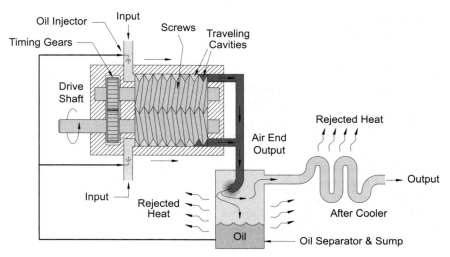

Figure 3-18   Screw compressor.

a housing. One of the screws is a left-hand thread and the other is a right-hand thread. The two screws are meshed together and synchronized with a set of timing gears. A motor is connected to a drive shaft, which rotates the screw set. The screws draw air in through the inputs, trapping it in the cavities that the screws make between the thread groove and the housing. The trapped air is carried to the discharge end via the traveling cavities, until it is forced out through the outputs.

The output air is fed through an air/oil separator, then through an after cooler and finally introduced into the system. The discharge temperature of a screw compressor must be controlled within a specific range, normally somewhere between 140 and 180°F. The reason for this is that the pump, which is referred to as the "air end," is lubricated by spraying oil mist into the air inputs. This oil has three functions—to lubricate, to provide a seal, and to provide some cooling. The reason that the output temperature must be regulated is that if the temperature is too high, there will be a high amount of oil contamination in the output air. If the output is too cool, then water vapor will start to condense into the compressor oil and the air end could be severely damaged. Most screw compressors are placed into a continuous duty application, therefore, they cannot rely on heat absorption to control their discharge temperature. Normally, they incorporate some type of after cooler, usually in the form of an air-cooled unit or an oversized receiver, the former being preferred.

When viewing a complete screw compressor, it can be little difficult to understand the system because of all the support equipment and components that surround the pump assembly. Figure 3-19 shows a typical pump/motor assembly that would be used in a screw compressor package. The motor can be mounted directly on the pump or the

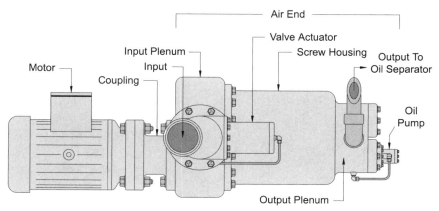

Figure 3-19   Motor/air end assembly.

assembly may utilize a vee belt drive. The input end of the pump has a plenum, which provides the incoming air access to the screw elements. The plenum carries an input valve and actuator, which is used to control the output of the compressor by throttling the input of the pump. The output plenum is on the opposite end of the pump and is connected to the oil separator.

Rotary Vane Compressors

The fifth type of compressor is the rotary vane compressor. These compressors are normally found as small units that provide air for an Original Equipment Manufacturer (OEM) application or a vacuum system. There are a few large rotary vane compressors, but they are not common. One of their big advantages is that they are very quiet. A small rotary vane compressor can be operated in the same room as a dentist chair and the patient will never hear it. Their biggest disadvantages are that they are only about 80 percent as efficient as a screw compressor; they are just as complex and have a shorter life expectancy.

Like the screw compressor, in theory a rotary vane compressor is a rather simple mechanism. Also like the screw compressor, the real world makes them a tad more complicated. Figure 3-20 shows a stylized sectional view of a rotary vane compressor. The pump consists of a cylinder with an internal rotor. The rotor is placed off-center so that its outside diameter is tangent with the cylinder's inside diameter. The rotor carries four vanes that are placed into slots. The vanes are usually spring-loaded and slide within the slots so that they are always in contact with the inside diameter of the cylinder. As the rotor vanes

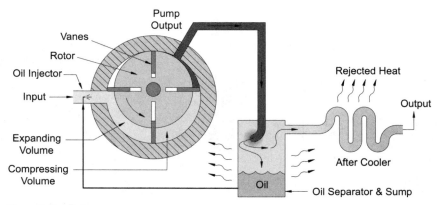

Figure 3-20   Rotary vane compressor.

rotate, an expanding volume is created which draws in air. As the rotor continues to rotate, the volume changes into a contracting cavity or compressing volume. The compressing volume continues to rotate and shrink until the air is exposed to the pump's output. The pump is lubricated and cooled in the same method that a screw compressor is, and has all the same requirements and support equipment.

As mentioned before, the biggest advantage of the rotary vane compressor is that they are very quiet units. This makes them popular choices for OEM applications, dental, and medical offices. They can also be found in tee shirt shops delivering air for the airbrushes. There are a number of companies that package systems specifically for these applications. Figure 3-21 shows a typical packaged dental compressor. These units will normally have a modest receiver with an encased pump mounted on top. The output is equipped with a master regulator, which is used to adjust the overall system pressure. Dental compressor packages may also incorporate some type of high-performance oil/water filter, air dryer, or both. When installing a system like this, it is important to

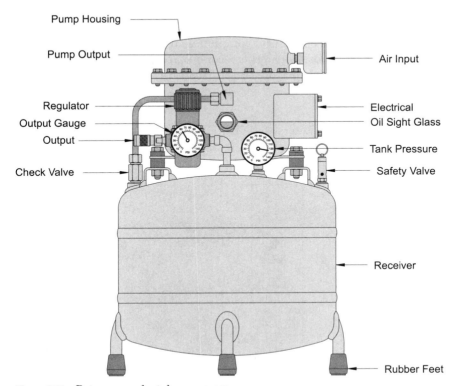

Figure 3-21   Rotary vane dental compressor.

consider the security of the compressor. These units are rather compact and can easily "walk away" if not properly protected.

Turbo Compressors

The sixth type of compressor is the turbo compressor. These compressors are normally found in applications that require very large volumes of compressed air such as large sandblasting operations, steel mills, automobile, and aircraft manufacturing plants. They produce compressed air at a very low cost-per-SCFM. This type of compressor is usually found in the 250+ hp range and I have seen turbo compressors as large as 5000 hp! One thing that really works against turbo compressors is that they only have a normal redundancy level. Because of their high capital cost, turbo compressors are rarely set up with a backup compressor. What this means is if a turbo compressor fails, a facility that may be worth a million dollars per hour comes to a complete halt. This is a risk that many large-plant engineers are unwilling to take. They will opt instead for a series of smaller screw compressors and trade efficiency for redundancy.

Figure 3-22 shows a stylized schematic view of a three-stage turbo compressor. The pumps are usually made up of three or more centrifugal compressors driven off of a common gearbox. The first stage compresses to a certain pressure, the second stage to the next higher, and the third stage to the output pressure. To control the heat of compression,

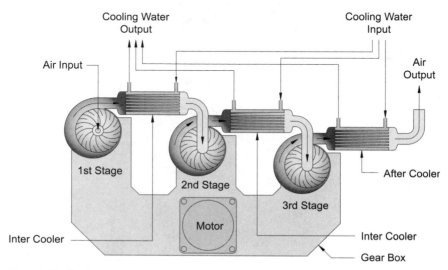

Figure 3-22   Turbo compressor.

most turbo compressors have water-cooled pumps that incorporate water-cooled intercoolers and have a water-cooled aftercooler.

Axial Compressors

A variation of the turbo compressor is the axial compressor. These compressors are usually specified for the largest applications. Their cost-per-SCFM is the lowest of all the compressor types. Axial compressors are usually over 1000 hp and have all the same drawbacks as the turbo compressor except on a more ambitious scale. The compressor itself is very much like the compressor stage found in most commercial jet engines.

Figure 3-23 shows a stylized schematic view of a five-stage axial compressor. The rotors are mounted to a common shaft. Each rotor is separated from the next by a stator. One rotor/stator set represents a single stage of the compressor. Because axial compressors cannot produce very high differential pressures, they typically have five to ten stages of compression. To handle the heat of compression, the case will incorporate a water jacket, which has a continuous flow of cooling water.

Injector Compressors

One other type of compressor, worth mentioning as a side bar only, is the injector compressor. These compressors are generally used for specialized process applications. However, there is at least one company manufacturing air compressors using this technology.

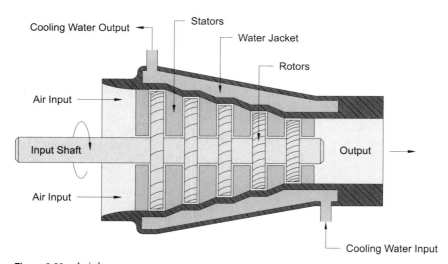

Figure 3-23   Axial compressor.

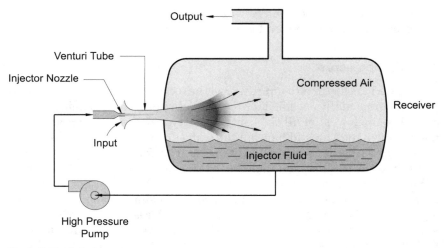

Figure 3-24   Injector compressor.

Figure 3-24 shows a schematic representation of an injector compressor. The heart of this type of compressor is the Venturi tube. The tube is mounted into the end of a receiver tank. A high-pressure fluid is injected through the Venturi tube. The fluid entrains air and compresses it into the receiver. The fluid and air separate after entering the receiver. The fluid falls to the bottom of the receiver and the compressed air output is at the top. The injector fluid is supplied to the injector via a high-pressure pump. The fluid is recycled from the bottom of the receiver and forms a closed loop system. The injector fluid can be almost any low-viscosity fluid, water being the most common.

Questions*

1. Name the six principal compressor types.

(A) Diaphragm, piston, two-stage piston, AC, rotary vane, and turbo
(B) Diaphragm, piston, two-stage piston, screw, rotary vane, and turbo
(C) Diaphragm, gas, two-stage piston, screw, rotary vane, and turbo
(D) Diaphragm, piston, automotive, screw, rotary vane, and turbo

2. What is the normal maximum pressure of a diaphragm compressor?

(A) 40 psi   (B) 60 psi   (C) 175 psi   (D) 125 psi

3. What is the most common small compressor?

(A) Turbo   (B) Rotary vane   (C) Piston   (D) Two-stage

4. What is the main reason for a two-stage compressor being more efficient than a single-stage unit?

(A) Motor size   (B) Intercooled heat removal   (C) Tank size
(D) Higher pressure

5. What is the general horsepower range of a two-stage compressor?

(A) 50–100 hp   (B) 1–5 hp   (C) 1000-hp+   (D) 2–30 hp

6. What is the normal tank pressure for a two-stage compressor?

(A) 175 psi   (B) 125 psi   (C) 60 psi   (D) 200 psi

7. Why is a higher operating pressure better than a lower one?

(A) Cooler air   (B) Less run time   (C) Higher air density
(D) Higher pressure

8. What compressor type dominates the compressed air industry in the 40- to 500-hp range?

(A) Piston   (B) Two-stage   (C) Turbo   (D) Screw compressors

9. Why are screw compressors referred to as "screw" compressors?

(A) The compression element is a pair of meshing screws.
(B) They screw to the ground when mounting.
(C) They fail frequently and as such screw their owners.
(D) It's just a nickname that the industry uses.

10. Why are screw compressors so common?

(A) Size
(B) Low purchase and operating costs

---

*Circle all that apply.

(C) Sound output

(D) Weight

11.   Why does the output temperature of a screw compressor need to be regulated?

(A) Efficiency

(B) To control oil discharge and to limit water contamination in the oil

(C) To keep the motor cool

(D) So that the paint doesn't burn

12.   What are some disadvantages of rotary vane compressors?

(A) Noise

(B) Purchase cost

(C) Oil consumption

(D) Poor efficiency and low life expectancy

13.   From an equipment standpoint, what does the rotary vane have in common with the screw compressor?

(A) Mounting   (B) Support equipment   (C) Water usage

(D) Screw speed

14.   What is the principal disadvantage of a turbo compressor

(A) Size   (B) Efficiency   (C) Energy usage

(D) Lack of redundancy

15.   How are most turbo compressors cooled?

(A) Air   (B) Water   (C) Oil   (D) Absorption

The Conservation of Mass and Energy as Applied to Compressed Air

To better understand air compression in general, it is important to have a basic knowledge of the role heat plays in this process. This role can best be described by reviewing the law of the conservation of mass and energy, as is applicable to air compression.

The law of the conservation of mass and energy simply states that the mass and energy of a standard cubic foot of air (SCFM) is constant, regardless of how the compressor may change its volume.

Two parameters change when we compress air, the first, and the one that we are most interested in, is pressure. If air is compressed to a smaller volume, then its pressure goes up inversely to the volume change. This means, if we compress air to one-tenth its original volume, then the pressure will increase 10 times. This is generally referred to as the "compression ratio." For example, if we compress air at atmospheric pressure (14.7 PSIA) to one-tenth its original volume, then the compression ratio will be 1:10 and the compressed pressure will be $14.7 \times 10 = 147$ PSIA. If we then expand this air to 10 times its compressed size, the compression ratio will be 10:1 and its expanded pressure will be $147 \div 10 = 14.7$ PSIA.

The second parameter, and the one that is the most troublesome, is temperature. Much like the pressure, the temperature also increases inversely to the volume change.

Therefore, if we compress air to one-tenth its original volume, then in a sense we also compress its temperature, which will increase 10 times its original temperature. Unfortunately, this temperature change is not so easy to calculate because the Fahrenheit and Centigrade scales are

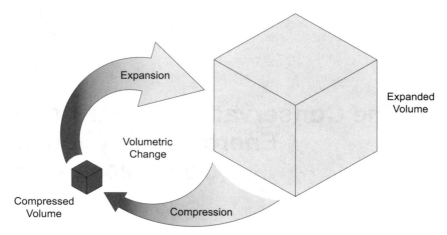

Figure 4-1 Expansion/compression graphic.

not absolute. To calculate the temperature in °F we must first convert to the Rankine (°R) scale, which requires adding 459.67 to the Fahrenheit reading. To calculate the temperature in °C we must first convert to the Kelvin (K) scale, which requires adding 273.15 to the Centigrade reading. As an example:

$$\text{Convert } 80°F \text{ to Rankine,} \quad 80°F + 459.67 = 539.67°R$$

$$\text{Convert } 27°C \text{ to kelvin,} \quad 27°C + 273.15 = 300.15 \text{ K.}$$

As an example of the change of temperature due to compression, if we compress air that has a temperature of 75°F, at a 1:10 compression ratio, then the temperature of the compressed air will be:

$$\text{For °F, } [(75°F + 459.67) \times 10] - 459.67 = 4887.03°F$$

$$\text{For °C, } [(24°C + 273.15) \times 10] - 273.15 = 2698.35°C.$$

Now anyone can tell you that the output of a 10-hp compressor is not 4887°F, after all, that would melt steel. The reason for this is that the real world interacts with the compression process and most of the energy is lost through the compressor heads, cylinders, piping, oil sump, crankcase, tank, and anything else that comes in contact with the process. All of these components are generally designed to extract as much heat as possible. Many compressors even use special coolers to aid in the extraction of heat. This is the reason for intercoolers on two-stage compressors and after coolers on screw compressors. What this translates to is the output temperature for most compressors is in the 125 to 300°F range. Reciprocating compressors rely heavily on heat absorption to control the output temperature. The machinery that makes up the compressor will absorb the heat of compression during the "on" cycle and then radiate it away during longer "off" cycles. This means a reciprocating compressor that runs continuously or has unusually long cycles will have a significantly higher discharge temperature than normal. A compressor in this type of duty will generally incorporate an after cooler to help remove the excess heat.

The most significant benefit of removing heat is that the more the heat is removed, the more efficient the compression process will be. This is because the hotter the air, the less dense it is. If you fill a tank with compressed air at 150°F, it will eventually cool down to match the ambient air that surrounds the tank. As the air cools, it condenses, and consequently, the pressure drops in the tank. For example, if you fill a tank to 125 psi at 160°F and then allow it to cool to 80°F, the pressure will drop to 60 psi! So you actually get half the work from your compressor!

Questions*

1.   In reference to the law of the conservation of mass and energy, what remains constant when compressing air?

   (A) Mass   (B) Temperature   (C) Pressure   (D) Energy

2.   What parameters change when air is compressed?

   (A) Color   (B) Flavor   (C) Pressure   (D) Temperature

3.   What is the compression ratio for 125 psi?

   (A) 1:6.12   (B) 1:8.50   (C) 1:10.20   (D) 1:11.90

4.   What would be the pressure for 1:5 compression ratio?

   (A) 44.1 psi   (B) 73.5 psi   (C) 161.7 psi   (D) 323.4 psi

5.   Convert 75°F to Rankine.

   (A) 534.67°R   (B) 584.67°R   (C) 296.15°R   (D) 354.15°R

6.   Convert 23°C to Kelvin

   (A) 534.67 K   (B) 584.67 K   (C) 296.15 K   (D) 354.15 K

7.   What is the typical output temperature range for commercial air compressors?

   (A) 100°F to 400°F   (B) 38°F to 90°F   (C) 125°F to 300°F
   (D) −40°F to 38°F

8.   What is the principal reason for removing heat from the compression process?

   (A) To prevent burning   (B) Efficiency   (C) To cool the motor
   (D) Oil life

9.   Where does most of the heat of compression go?

   (A) Pump oil
   (B) Applications
   (C) Compression system components
   (D) Tank

---

*Circle all that apply.

5

The Compressed Air Supply

The fundamentals of a compressed air system are rather simple to understand. The basic system will typically consist of a pump, electric motor, pressure switch, and an air tank or "receiver." When the motor is energized, it drives the pump, which draws in ambient air and pumps it into the receiver.

Figure 5-1 shows a schematic of an ordinary packaged reciprocating compressor. The pump is driven by the electric motor, both of which are mounted on top of the receiver. When the pressure in the receiver reaches a predetermined upper limit, the pressure switch turns off the motor. When the pressure in the receiver reaches a predetermined lower limit, the pressure switch turns on the motor.

In addition to controlling the motor state, the pressure switch also has an unloader function. When the upper pressure limit is reached, the switch turns off the motor and opens a vent valve, which is connected to the output of the pump. This venting provides a zero load condition when the motor restarts for the next cycle. A check valve on the output of the pump, between the unloader port and the receiver, prevents the pressure in the receiver from venting through the unloader valve during off periods. For compressors that have short cycle intervals, the system may be set up to run continuously. In this case the motor is not turned off, just the unloader port is opened. This is done because starting and stopping an electric motor too often may cause it to overheat and consequently burn up.

Screw Compressors

Compressed air systems larger than 30 hp are typically pumped by a screw compressor. These systems normally have a separate receiver, as shown in Fig. 5-2. Most screw compressors are continuous run systems

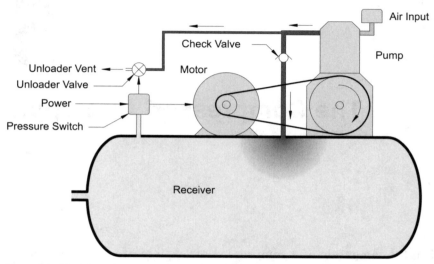

Figure 5-1   Reciprocating compressor schematic.

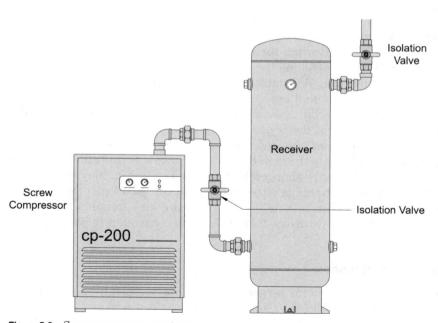

Figure 5-2   Screw compressor system.

and do not unload when maximum pressure is reached. The input of the pump, or "air end," has a throttling valve, which controls the amount of air that can be drawn in. It should be noted that some of these compressors are now being delivered with variable frequency motor drives, which are utilized to control the speed of the motor and, consequently, the output of the air end.

However, it should be noted that, there are some small packaged screw compressors, which operate in the same "unloader" fashion as the packaged reciprocating compressor shown in Fig. 5-1. These systems are almost always less than 30 hp and are marketed to compete with their reciprocating counterparts.

Duplex Compressors

Duplex compressors are rather common systems for applications that have high peak load demands or systems that require a high degree of redundancy. These installations may be either a single unit with two pump/motor assemblies, as shown in Fig. 5-3, or two completely separate compressors with common controls, as shown in Figs. 5-4 and 5-5.

The pumps on a duplex compressor can run separately or together. Normally, the pumps toggle during operation. That is to say that one pump runs for a cycle and then the opposite pump runs for the next cycle. Duplex compressors also operate as peak load systems. If the tank

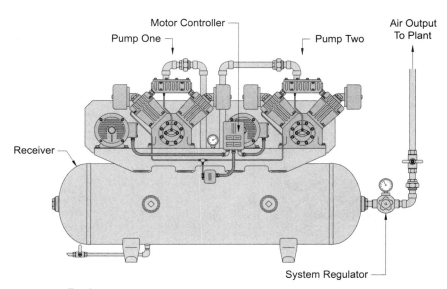

Figure 5-3 Duplex compressor.

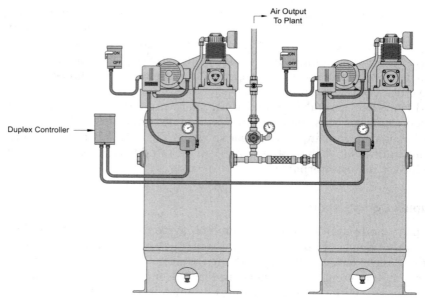

Figure 5-4   Duplex installation w/two vertical compressors.

pressure drops below a predetermined lower limit, the nonoperating pump will turn on and both pumps will run until the system pressure is reached.

Figure 5-3 shows a typical packaged duplex compressor. The unit has two pumps mounted on a common receiver. Units like this will be equipped with an appropriate controller mounted to the frame.

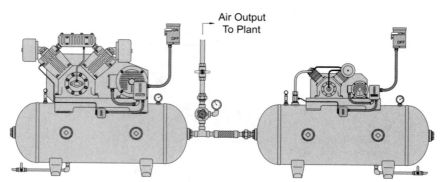

Figure 5-5   Duplex installation for peak load applications.

Redundant Compressors

Screw compressor systems are normally set up for redundancy only. Most systems are operated on one compressor for a week and then switched to the opposite unit for the next week. This arrangement allows each compressor to be serviced without disturbing the compressed air supply to the plant. Additionally, a reserve compressor is always available in the event the other unit fails unexpectedly. Figure 5-6 shows a typical redundant screw compressor setup.

Duplex Controllers

Duplex controllers are generally delivered as an integral part of a packaged duplex compressor. However, if two independent compressors are set up as a duplex system, then a stand-alone controller must be added, as shown in Fig. 5-4. These controllers will toggle the compressors during operation. That is to say that one compressor will run for the first

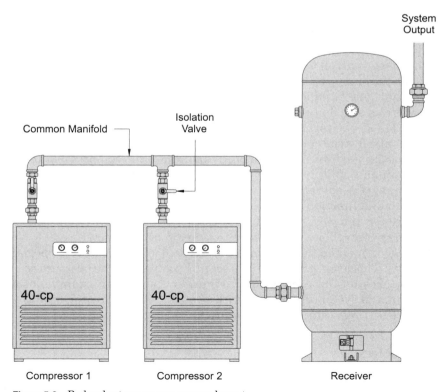

Figure 5-6   Redundant screw compressor layout.

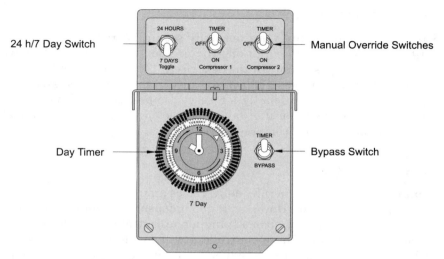

Figure 5-7   24 h/7 day duplex controller.

cycle and the opposite will run for the next cycle. Most duplex controllers will also allow the system to operate in peak demand service as
well. These controllers are normally equipped with manual override
controls that allow turning on and off one or the other compressor.

Another type of duplex controller (Fig. 5-7) is intended to toggle the
compressors on a daily or weekly basis. These controllers are little more
than a 24 h/7 day toggle timer with a control relay that interrupts the
motor controller coils on the two different compressors. Some controllers
are also equipped with a 7 day timer that allows the compressors to be
turned on and off in reference to the hours and days of the week that
the company is operating. In general, they are also equipped with a
peak demand function.

Figure 5-8 shows a typical duplex screw compression system with a
24 h/7 day controller that has been set up for redundancy.

The peak demand duplex control scenario does not require a special
controller. This arrangement can be set up by adjusting the pressure control switches on the compressors. Figure 5-9 shows an example of a
peak demand system configured with two 2-stage compressors and the
pressure settings that are appropriate for this application.

After Coolers

An after cooler is generally specified, when a compressor is producing
unacceptably high output temperatures. There are a number of reasons

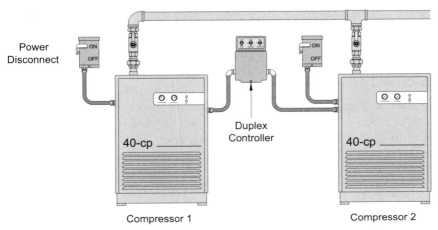

Power
Disconnect

Duplex
Controller

40-cp

40-cp

Compressor 1

Compressor 2

Figure 5-8 Redundant screw system.

why it might be necessary to cool the discharge of a compressor. In the case of a reciprocating unit that is running continuously, the output temperature will be much higher than normal and must be cooled externally. Chlorofluorocarbons (CFC)-based dryers generally require an input temperature of no higher than 100°F, and the output of a compressor is rarely that low. In southern latitudes, compressors are forced to operate in high ambient temperatures and require additional cooling. It should be noted that continuous-run screw compressors are almost

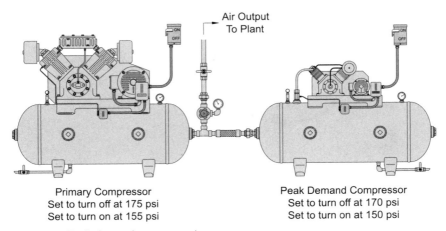

Air Output
To Plant

Primary Compressor
Set to turn off at 175 psi
Set to turn on at 155 psi

Peak Demand Compressor
Set to turn off at 170 psi
Set to turn on at 150 psi

Figure 5-9 Peak demand system settings.

always equipped with an after cooler from the factory. Even so, these factory-equipped units are intended for average conditions and an additional after cooler may be required for some situations.

Air-Cooled After Coolers

Air-cooled after coolers (Fig. 5-10) are simply radiators that can handle high internal pressures. They are typically cooled by an integral electric fan that forces ambient air across their fins. This type of after cooler is very common and is a standard component on most screw compressors and many reciprocating compressors. If applied properly, they can be very effective in removing excess heat in certain situations. The significant drawback of air-cooled after coolers is that they rely on the ambient temperature. Their normal cooling capabilities can only provide output temperatures of approximately 20 to 50°F above ambient. This temperature spread is generally referred to as the approach temperature.

Figure 5-11 shows a typical installation of an after cooler on a packaged reciprocating compressor. The after cooler is usually mounted on

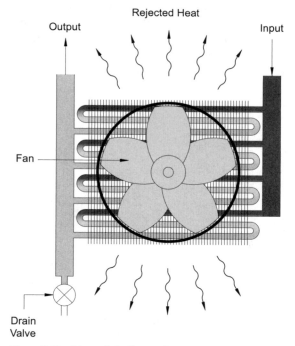

Figure 5-10   Air-cooled after cooler.

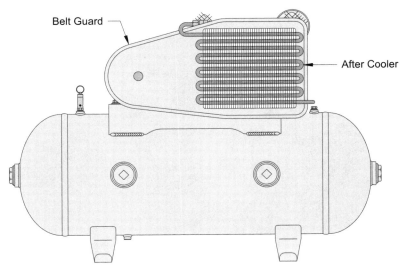

Belt Guard

After Cooler

Figure 5-11   Reciprocating compressor with after cooler.

the belt guard and the pumps flywheel acts as the fan. Integral after coolers are offered as an option on most reciprocating compressor packages and are a strongly recommended accessory. Compressors that are set up for continuous-run operation, or have unusually high duty cycles, should be equipped with an after cooler.

Smaller compressors that are not equipped with an integral after cooler can be set up with a stand-alone unit that is sized to match their specific requirements. These units are typically inexpensive and can provide a substantial performance improvement to your compression system. When placing these units, great care should be taken to ensure that ample airflow and ventilation are available. It is a good idea to equip the output of the after cooler with a water trap, as shown in Fig. 5-12.

Figure 5-13 shows a typical commercial after cooler, which is configured for a large compression system. These units are generally placed outside and adjacent to the compressor room. They should always be installed in an open area, which ensures ample airflow and ventilation. As with their smaller counterparts, it is a good idea to equip the output of the after cooler with a water trap as shown.

Water-Cooled After Coolers

The second type of after cooler is water cooled. These are usually a tube-in-shell heat exchanger. The output of the compressor is directed through the tubes and the shell is flooded with a continuous flow of

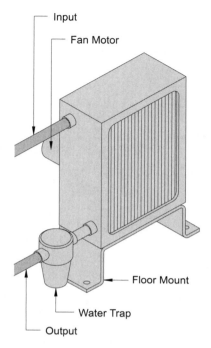

Input

Fan Motor

Floor Mount

Water Trap

Output

Figure 5-12   Stand-alone after cooler.

water. Water-cooled units have two distinct advantages. First, they are independent of ambient air temperatures, so they are applicable in situations where the outside environment may not be suitable for an air-cooled unit. Second, they are very compact and are ideal for cramped compressor rooms. Their drawback is that they require a substantial amount of cooling water. If sized correctly, they can approach temperatures as low as 5°F in reference to the cooling water. Figure 5-14 shows a schematic representation of a water-cooled after cooler. Figure 5-15 shows how a commercial heat exchanger is set up to act as an after cooler. Pay special attention to the trap section. As the compressed air is cooled, water will condense out and collect on the internals of the after cooler. This condensed water must be trapped and periodically drained.

Air Dryers

Air dryers are fairly common accessories for compression systems, the most common being CFC-based refrigerated dryers. Figure 5-16 shows a typical installation on a packaged reciprocating compressor. Figure 5-17 shows a typical installation on a screw compression system. The manifold

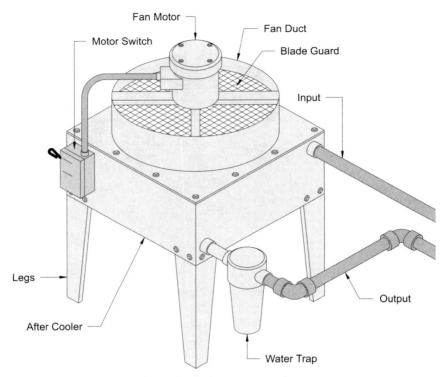

Figure 5-13 Stand-alone after cooler for large compression systems.

and valve set between the receiver and dryer in Fig. 5-16, and the dryer in Fig. 5-17, are bypass manifolds. These manifolds are required so that the dryer may be taken off-line for servicing. A description of how refrigerated dryers operate is given in Chap. 6.

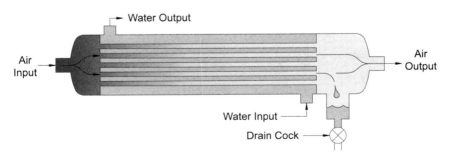

Figure 5-14 Water-cooled after cooler schematic.

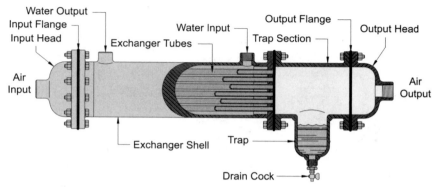

Figure 5-15   Water-cooled after cooler.

Instrument Air

Refineries, chemical plants, semiconductor manufacturing facilities, food processing installations, and the like typically specify instrumentation quality air for their pneumatic control systems. Instrument air is generally specified at −40°F dew point. Many of these systems also incorporate a utility air output, which generally does not carry a dew point specification. The utility outputs produce notoriously poor air quality and, in many cases, even the instrument air is of poor quality due to a lack of maintenance or inadequate design of the compression package.

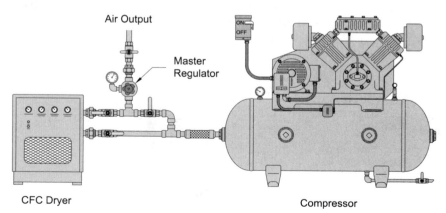

Figure 5-16   Packaged reciprocating compressor w/CFC dryer.

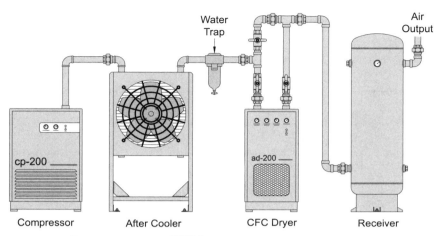

Water Trap

Air Output

cp-200

ad-200

Compressor   After Cooler   CFC Dryer   Receiver

Figure 5-17 Screw compressor w/CFC dryer.

Figure 5-18 shows a typical instrument air system. These systems normally consist of a compressor(s), a wet tank (which also serves as the utility air receiver, after cooler and water separator), a regenerative "twin tower" desiccant dryer, a dry tank, and an output filter. The drain discharges should be plumbed to an oil/water separator.

Systems laid out in this manner have several significant drawbacks. Because the wet receiver has very little effect on the discharge temperature of the compressor, it can only separate gross water and oil contamination, forcing the desiccant dryer to act as a primary dryer and to carry nearly the entire water and oil load from the compressor discharge. This places the desiccant charge in a situation it is not actually designed to handle, and that will significantly reduce its effective life. Most users set up an aggressive regeneration cycle in a vain attempt to compensate for this overload condition, but this is only a band-aid applied to a far more significant problem.

Oil contamination is particularly detrimental to the desiccant charge. Oil-contaminated desiccant cannot be regenerated and eventually will become completely saturated with oil, rendering the dryer ineffective and forcing the user to replace the charge. Additionally, and especially in southern climates, the temperature that the dryer receives is oftentimes too high. Realistically, a desiccant dryer should see an input temperature well below 100°F and ideally below 70°F. Temperatures any higher will dramatically reduce the performance of the dryer. If the input temperature is too high, then the dryer cannot regenerate as quickly as it absorbs water and the system is rendered ineffective. The

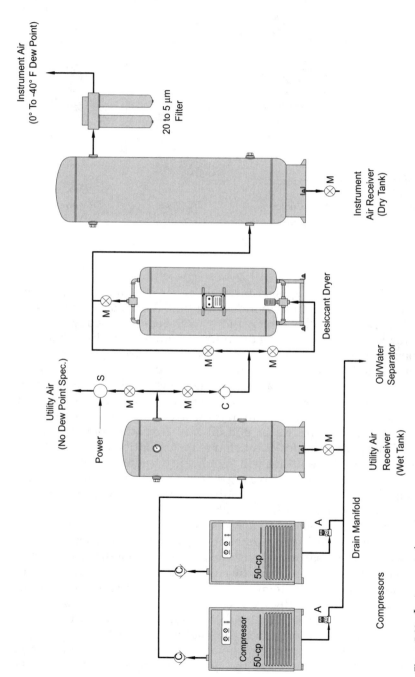

Figure 5-18 Instrument air system.

net effect is that the discharge has a water content that is significantly higher than the required specification.

To avoid excessive temperatures, overloading, and oil contamination, the desiccant dryer should be used only as a secondary dryer. The primary dryer may be in the form of a water-cooled after cooler or a refrigerated dryer, the latter being the preferred choice.

One solution is to include a CFC-based refrigerated dryer as a first stage. However, CFC-based dryers are also dependent on ambient temperatures and typically require input temperatures of no greater than 100°F. The solution is to use an after cooler as a precooling unit before the CFC-based dryer. It should also be noted that there are "high temperature" CFC-based dryers that will accept input temperatures as high as 150°F. These units are considerably more expensive than their "low temperature" counterparts. CFC-based dryers also have other drawbacks. They are rather complex and delicate mechanisms with a short life expectancy, usually in the 5- to 8-year range. Additionally, they do not fare well in the type of environments where compression systems are typically installed.

The best solution is to utilize an Elliott cycle refrigerated dryer as a first stage. Ambient temperatures have no effect on the operation of these dryers and they can accept input temperatures as high as 365°F. These dryers have the added advantage of operating between the compressor output and the utility receiver, which translates to high quality utility air in addition to removing the burden on the desiccant dryer.

Figure 5-19 shows a typical instrument air package with an Elliott cycle refrigerated dryer in the dual role of an after cooler and first-stage dryer.

Field Compressors

There are quite a number of applications where engine-driven compressors are used. These uses range from contractor compressors in the 2- to 5-hp range to large mining operations, which may use skid-mounted compression systems driven by 1000 hp diesel engines. Most of us have seen the compressors mounted in the back of pick-up trucks for field servicing of tractor-trailer tires. We have also seen road crews using trailer mounted construction compressors to provide air for the jackhammers. These are just two of the more visible uses for field compressors.

Field reciprocating compressors (Fig. 5-20) operate in much the same manner as a fixed unit, except that the gasoline engine runs continuously. When the receiver reaches its high pressure, a special continuous-run unloader valve vents the output of the pump to atmosphere and

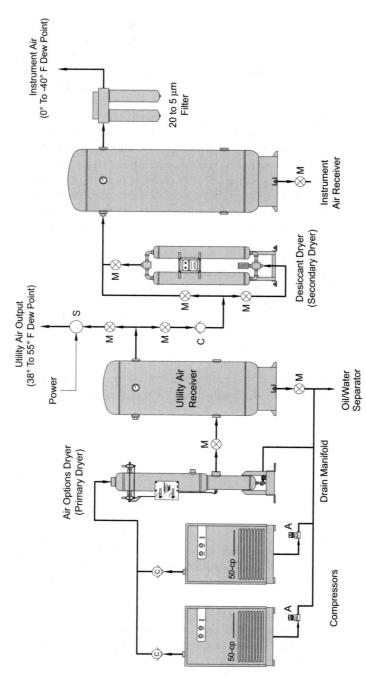

Figure 5-19 Improved instrument air system.

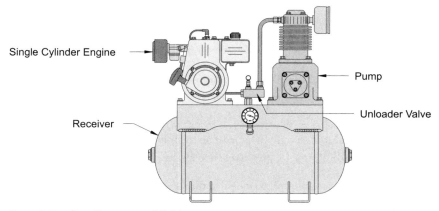

Figure 5-20   Gasoline-powered field compressor.

lowers the throttle of the engine down to idle speed. When the receiver pressure drops to the lower limit, the unloader reconnects the output of the pump and opens the throttle to run speed.

Larger diesel compressors, as shown in Fig. 5-21, are typically screw compressors that operate in the same way as a fixed unit.

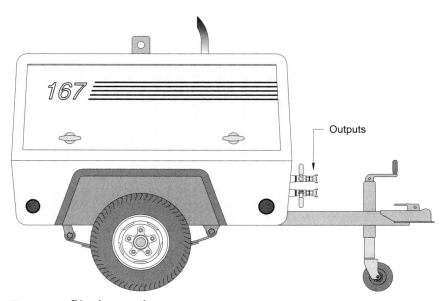

Figure 5-21   Diesel-powered screw compressor.

Questions*

1. What additional function does the pressure switch usually have?

(A) Unloading   (B) Alarm   (C) Timer   (D) Motor controller

2. Why is it important to unload the compressor pump when it turns off?

(A) Reduce heat   (B) Save electricity   (C) Lower starting loads
(D) Reduce noise

3. What component is usually separate on systems larger than 30 hp?

(A) Motor   (B) Pump   (C) Pressure switch   (D) Receiver

4. For what two system applications are duplex compressors normally installed?

(A) Redundancy   (B) Food service   (C) Peak demand
(D) Auto repair

5. For what time interval are redundant screw compressors typically operated at?

(A) 1 h   (B) 30 min   (C) 24 h   (D) 7 days

6. What is the most common type of dryer?

(A) Desiccant   (B) Refrigerated   (C) Membrane   (D) Absorption

7. What dew point is instrument air generally specified at?

(A) −40°F   (B) 40°F   (C) 38°F   (D) 0°F

---

*Circle all that apply.

6

Compressed Air Dryers

Water contamination is one of the most important problems to address when dealing with compressed air. However, it is an area that is usually neglected. Water condensation, buildup, and removal are serious problems within compressed air systems, causing millions of dollars in damage annually to these systems and the equipment they serve. Water can accelerate rusting in the receiver and piping, it will flush out lubricating oils in delicate air tools, gum up seals in valves and cylinders, fog spray painted finishes, clump sandblasting media, and build up in automobile tires. To combat this rather significant problem there are a wide variety of devices on the market.

So how does the water get into the compression system? We must go back to the temperature problem reviewed in Chap. 4 to better understand this question. Water content in air is generally described in percent of relative humidity. If the air is at 50 percent relative humidity, then it has half of the total water vapor that it can hold at its current temperature. However, it is important to understand that temperature plays a critical role in this arena. The higher the air temperature, the higher the air's affinity for water vapor will be. As an example, air that has a temperature of 50°F and a relative humidity of 100 percent has a much lower water content than air that has a temperature of 100°F and a relative humidity of 50 percent. With this in mind, consider what happens when we compress air. The pump draws in air at, say, 75°F and 60 percent relative humidity. During the compression process, the air volume is reduced and its pressure and temperature increase. Pressure has a minimal effect on air's ability to carry water vapor; on the other hand temperature has a profound effect. The compressed, high-temperature air is saturated and may be as much as 25 percent water by volume.

Figure 6-1 Are there times when it feels like your air tools are operating off of a garden hose?

As this hot, saturated, compressed air comes in contact with the various components of the system, it starts to cool. As the air cools, its affinity for water vapor lowers and it starts to shed its water content to match its temperature. The water that is shed condenses onto the inside of the tank and pipes. You now have liquid water contaminating the inside of your compression system. For most compressor users, installing water traps this is the first thing they try when attempting to control water build-up.

Traps

Traps can be effective if applied within the context of a properly designed plumbing system; however, for most users these devices provide no benefit whatsoever.

A drop trap (Fig. 6-2) is simply a pot with a separator baffle that forces the air to exit the bottom and make a 180° turn back up to the output. The water drips into the bottom of the housing and is periodically drained through the drain cock. Many drop traps are built "in house" by their users.

Figure 6-3 shows a typical drop trap. The body is a piece of 3-in. standard wall pipe with weld caps on both ends. A half-coupling is welded on the bottom cap to provide a drain port. The outputs are simply two or three half-couplings welded to the top of the body. The input and baffle is a piece of $^3/_4$-in. schedule 80 pipe that has been threaded on one end and welded in place as shown.

Figure 6-4 shows a typical drop trap, which has been constructed from ordinary pipe fittings. The body is a 2-in. standard wall pipe nipple with a lateral tee on the top and a reducer on the bottom. The lateral tee acts as a baffle. The reducer is fitted with a bushing and drain cock. The lateral tee is capped with a bushing that matches the overhead feed. The branch of the tee is equipped with a bushing and a 45° street elbow.

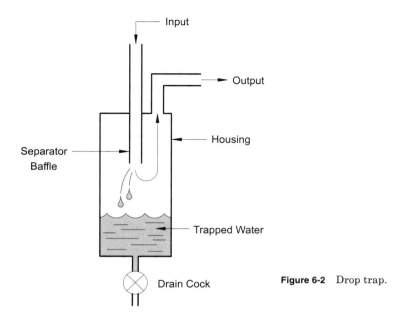

Input

Output

Housing

Separator
Baffle

Trapped Water

Drain Cock

Figure 6-2 Drop trap.

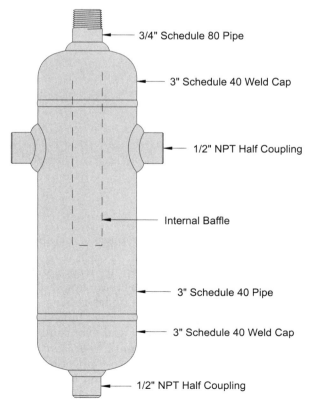

3/4" Schedule 80 Pipe

3" Schedule 40 Weld Cap

1/2" NPT Half Coupling

Internal Baffle

3" Schedule 40 Pipe

3" Schedule 40 Weld Cap

1/2" NPT Half Coupling

Figure 6-3 Welded drop trap.

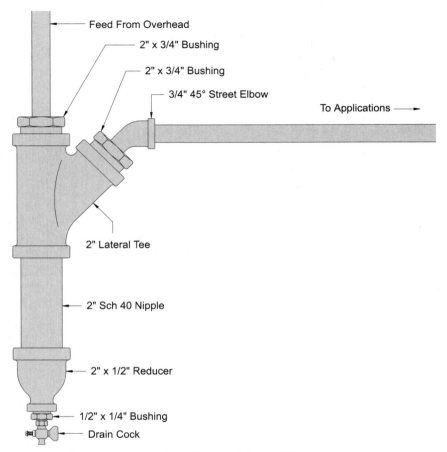

Feed From Overhead

2" x 3/4" Bushing

2" x 3/4" Bushing

3/4" 45° Street Elbow

To Applications ⟶

2" Lateral Tee

2" Sch 40 Nipple

2" x 1/2" Reducer

1/2" x 1/4" Bushing

Drain Cock

Figure 6-4   Drop trap construction with threaded pipe fittings.

Figure 6-5 shows an automatic drain installed in place of a standard drain cock.

In many instances, large distribution systems require a second level of water separation. The long pipes may sag, which create low sections. Further, it may be necessary to route pipes through undesirable geometries, such as a loop up above an overhead door or a run of buried pipe. These situations create areas that water can pool and eventually the pipe will fill with water. Figure 6-6 shows a schematic representation of a typical line trap. The trap is simply a pot with two connections and an internal baffle. The airflow is forced to flow down and around the bottom of the baffle and any liquid water in the stream collects in the trap. The trapped water is periodically drained through the drain valve.

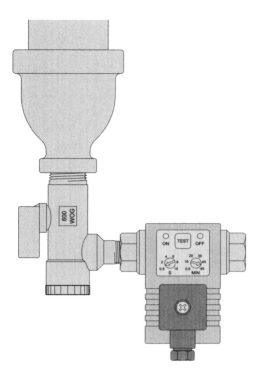

Figure 6-5 Automatic electronic drain.

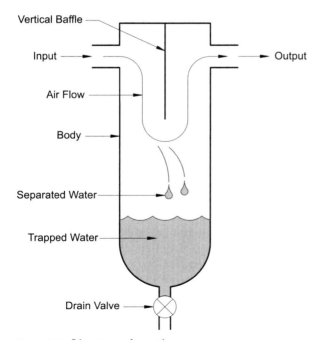

Vertical Baffle

Input →

Air Flow →

Body →

Separated Water

Trapped Water

→ Output

Drain Valve →

Figure 6-6 Line trap schematic.

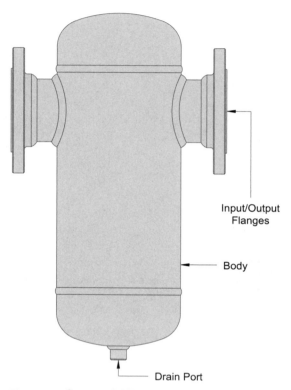

Input/Output
Flanges

Body

Drain Port

Figure 6-7   Commercial line trap.

Commercial line traps are generally a welded assembly as shown in Fig. 6-7. These types of traps are extremely durable pieces of equipment and, when installed properly, will provide decades of service.

As with drop traps, line traps are only effective in the context of a properly designed compressed air distribution system.

A variation of the line trap is the cyclone trap (Fig. 6-8). These are the single most common trap on the market today. They are very inexpensive and can be purchased singularly or as an integral part of a regulator or lubricator. The principle behind these traps is that they force the air to flow in a circular pattern and centrifugally separate the water. In theory, they seem like a good idea; in practice, they're not. These devices rely heavily on high airflow for separation. Unfortunately, they are usually specified for applications that do not have a high enough airflow to make them effective.

Commercial cyclone traps (Fig. 6-9) generally incorporate a removable bowl that is held in place with a locking collar and sealed with an O-ring. These units usually need regular cleaning, so a good supply of replacement

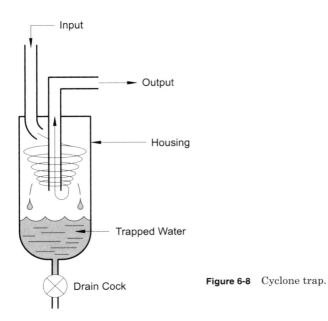

Input

Output

Housing

Trapped Water

Drain Cock

Figure 6-8 Cyclone trap.

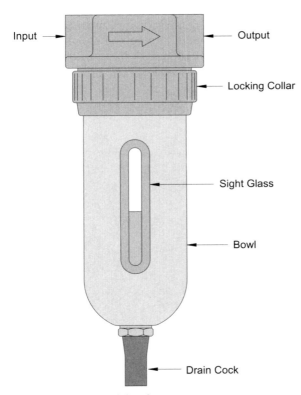

Input

Output

Locking Collar

Sight Glass

Bowl

Drain Cock

Figure 6-9 Commercial cyclone trap.

O-rings should be kept on hand. Opening up one of these units will frequently damage the O-ring, especially if it has been in service for a long time. The bowl is typically equipped with a "push to vent" drain cock. These valves are generally of very poor quality and, oftentimes, will start to leak after only a few actuations. When selecting a trap, the old adage applies: "you get what you pay for."

Plate Separators

Taking the trap a little further is the plate separator. These are comprised of a series of elements that force the compressed air to flow through a labyrinth of channels, which trap the water when it comes in contact with the internals. The elements are designed to separate and drain the water into the bottom of their housings.

Just as common as the cyclone trap is the plate separator (Fig. 6-10). It is simply a stack of separator plates that have spaces between them so that air can flow through. Much like the cyclone trap, these units are very inexpensive and generally provide rather poor performance.

Commercial plate separators (Fig. 6-11) are oftentimes indistinguishable from a cyclone trap. They generally incorporate all the same attributes and are usually constructed using the same outward components. Like the cyclone trap, it's a good idea to maintain a good supply of replacement O-rings for the bowl seal.

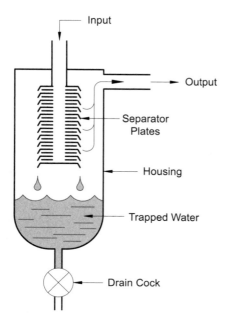

Input

Output

Separator Plates

Housing

Trapped Water

Drain Cock

Figure 6-10 Plate separator.

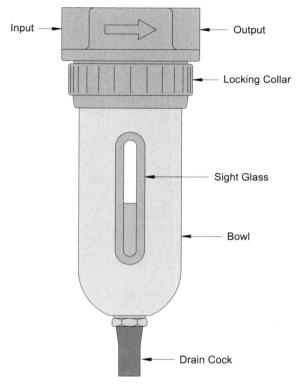

Input —

Output

Locking Collar

Sight Glass

Bowl

Drain Cock

Figure 6-11 Commercial plate separator.

Coalescing Filters

Taking the plate separator a little further is the coalescing filter. These are generally comprised of hundreds of closely placed elements that force the compressed air to flow through restrictive gaps in the stack. Like the plate separator, the water separates when it comes in contact with the elements, except that with the coalescing filter this separation is considerably more efficient.

The most effective coalescing filter is the micro grid (Fig. 6-12). These units are similar to plate separators, except the plates are spaced very close to one another. Close enough that micro quantities of water will wet the plate surfaces and be separated. These types of filters are fairly effective, but have the drawback of being rather expensive to purchase and fragile in operation. They are susceptible to oil contamination and generally require teardown and cleaning on a regular basis.

Figure 6-13 shows a typical plate array coalescing element. The plates are stacked intermittently with spacers so as to provide gaps for the air

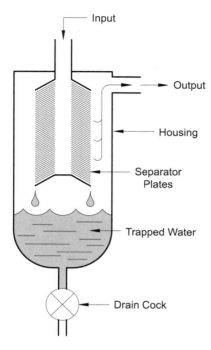

Figure 6-12 Micro grid coalescing filter.

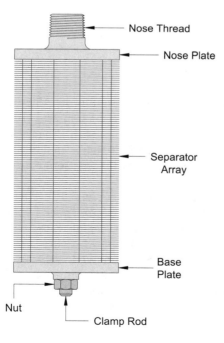

Figure 6-13 Stacked plate element.

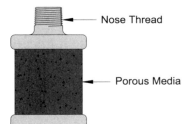

Nose Thread

Porous Media

Figure 6-14 Porous media element.

to flow through. The plate stack is usually clamped between a nose plate and base plate. Figure 6-10 shows a porous media coalescing element. The media is generally fixed permanently to the plates. These elements are not serviceable and are typically replaced when cleaning is no longer effective.

Porous media coalescing filters can be quite effective if applied properly. Figure 6-14 shows how one of these filters would be set up at the application site. The system air should be supplied from some type of first-stage dryer, such as a refrigerated unit as discussed later in this chapter. Figure 6-15 shows how a stacked plate filter would be used in a systems role. This arrangement is generally suitable for low-performance applications, if the host compressor is equipped with an aftercooler and has a fairly low-duty cycle as shown in Fig. 6-16.

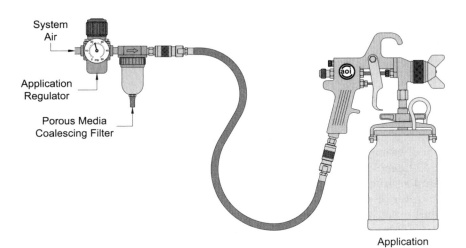

System
Air

Application
Regulator

Porous Media
Coalescing Filter

Application

Figure 6-15 Coalescing filter installation.

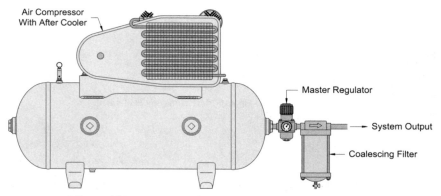

Air Compressor
With After Cooler

Master Regulator

System Output

Coalescing Filter

Figure 6-16   Micro grid in a systems role.

Water-Cooled Aftercoolers

Aftercoolers are generally associated with removing excess heat from the compression system. However, in certain situations, they can be very effective in the removal of water. If a chilled water source is available, they can be set up to cool the compressed air and condense out the water vapor. A cooled water source can be well water, a stream, a chilled water system, the water that a ship is floating in, or even spring run off. Generally speaking, the water temperature should be lower than 50°F. The heat exchanger must be sized to match the heat load of the compressor versus the temperature and flow rate of the chilled water source. Additionally, the heat exchange should be equipped with a water trap. The drawback of this type of system is that the aftercooler will require a rather substantial amount of water flow. On the plus side, these devices are very compact, have no moving parts, and are exceptionally reliable. Figure 6-17 shows a typical water-cooled aftercooler that has been equipped for drying applications.

CFC-Based Refrigerated Dryers

The most common "effective" dryer is the refrigerated dryer. This type of dryer is really like a supercharged aftercooler. Instead of relying on a chilled water source for cooling, this type of dryer uses a refrigeration cycle, the most common being the CFC cycle. In the past, dryers have been manufactured using an ammonia cycle, but it is unlikely that you will ever encounter one of these units. A third type of refrigerated dryer uses the Elliott cycle, which is discussed in further detail in this chapter.

A CFC-based refrigerated compressed air dryer (Fig. 6-18) removes water by cooling the compressed air and condensing the water vapor out in a controlled environment. These units consist of a heat exchanger that

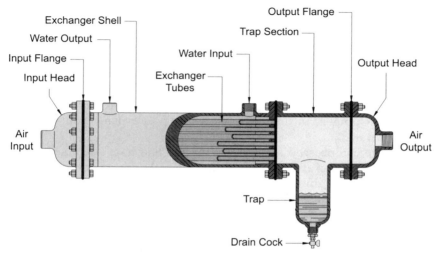

Figure 6-17   Water-cooled aftercooler.

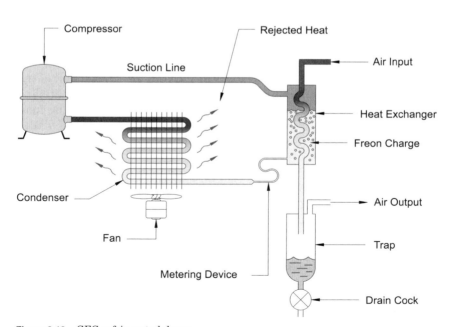

Figure 6-18   CFC refrigerated dryer.

is similar to a water-cooled aftercooler. Instead of using water as the coolant, liquid CFC fills the shell of the heat exchanger. The liquid CFC is maintained at a pressure that allows it to boil at 38°F.

After the CFC boils, the vapor is drawn through the suction line into a compressor, which compresses the CFC to a high pressure and high temperature. The high-pressure/temperature CFC is cooled in the condenser and relaxes into its liquid state. The liquid is reintroduced into the heat exchanger via the metering device and a closed refrigeration cycle is formed.

When the compressed air passes through the heat exchanger, it is cooled to the temperature of the boiling CFC. As the compressed air is cooled, it loses its ability to retain moisture and the water vapor condenses onto the inside of the exchanger tube.

This type of dryer does have a few significant drawbacks. Their input temperature is generally limited to 100°F, although special high-temperature versions are available. The high-temperature versions have an added air-cooled precooler on their input. Generally speaking, these units are considerably more expensive than their low-temperature counterparts. Another significant drawback of CFC-based dryers is that they have a short life expectancy, generally in the 5- to 8-year range. They are also rather complex and delicate pieces of equipment that don't really fare well in the type of environments that compressors usually operate in. Lastly, you must use a licensed CFC technician to perform service on these units. Most companies do not have a CFC-qualified technician on the payroll, so an outside vendor must be called in for servicing.

Higher-performance dryers will oftentimes incorporate an air reheater to improve efficiency. The concept is to reheat the cooled air from the output of the dryer and therefore expand the volume. The output of the dryer is passed through an air-to-air heat exchanger that is heated by the incoming air. This also has the effect of passing some of the heat energy directly to the output, which mitigates the thermal load placed on the refrigeration system. Figure 6-19 shows a schematic representation of a reheater.

High-temperature CFC dryers are typically a standard dryer design with an air-cooled precooler added to remove some of the heat associated with the incoming air. The same effect can be accomplished by simply adding an aftercooler to the output of the compressor and by using a standard dryer. However, high-temperature dryers provide a much more compact installation. Figure 6-20 shows a schematic of a high-temperature CFC-based dryer.

CFC-based dryers are generally delivered packaged in a single cabinet as shown in Fig. 6-21. These units typically carry a control panel and gauge set that show input temperature, air pressure, CFC high pressure, and CFC low pressure. The cabinets are vented and carry a forced air

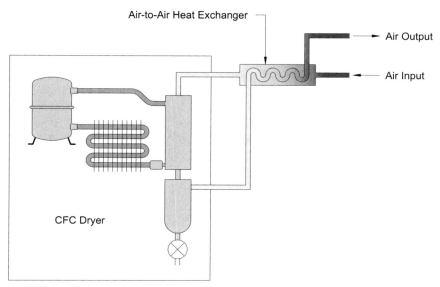

Figure 6-19   Reheated output.

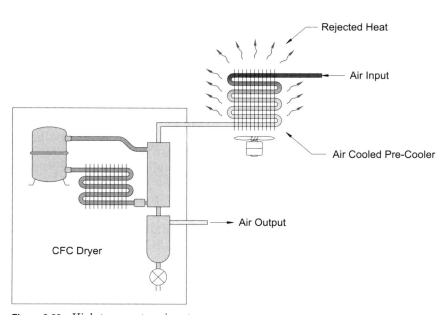

Figure 6-20   High-temperature input.

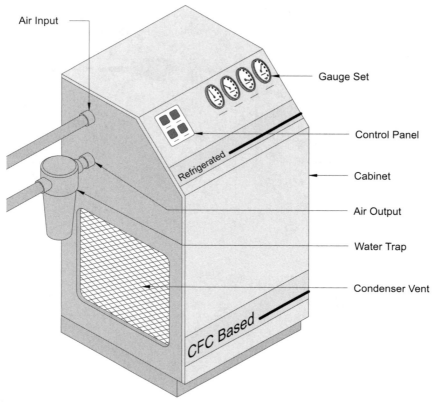

Air Input

Gauge Set

Control Panel

Refrigerated

Cabinet

Air Output

Water Trap

Condenser Vent

CFC Based

Figure 6-21   Commercial refrigerated dryer.

system to cool the condenser coil. The output of the unit may have an external water trap, as shown, or the trap may be internal. Dryers up to 200 standard cubic foot per minute (SCFM) are typically set up to operate on 120/220 VAC, dryers over 200 SCFM are usually 220/440 VAC. It is imperative that these units are installed in a well-ventilated area. Restricting airflow or high-ambient temperatures will greatly reduce the performance of these systems. The condenser coil should be cleaned at least once a quarter, and once a month in particularly dirty environments. The refrigeration system should be serviced every two years of operation. Even with regular service these units normally have a life expectancy of only five to ten years.

Another variation of the CFC-based dryer is the cycling dryer. Instead of having a CFC/air heat exchanger, these units carry a large thermal mass, usually in the form of a volume of water. The refrigeration system cools the water and, in turn, the water cools the compressed air. When the charge of water gets down to the operational temperature, a thermostat turns off the refrigeration. When the water temperature climbs above a preset temperature, the thermostat turns on the refrigeration. Cycling dryers are an excellent choice if your system has a wide variation of airflow. Larger dryers, in excess of 2000 SCFM, are normally designed to cycle. Additionally, larger dryers generally carry redundant refrigeration systems that can be manipulated to match the heat load for a given flow rate.

Figure 6-22 shows a schematic representation of a cycling dryer. On smaller units, the heat exchanger is usually an insulated bucket with two copper coils immersed in a water bath. The refrigeration system uses one coil, while compressed air is passed through the other. Figure 6-23 shows a high-capacity cycling dryer.

When installing a CFC-based air dryer with a packaged compressor, it is important to consider the peak flow rate that the unit will be expected to carry. Normally the dryer is installed on the output of the receiver and, therefore, it can be subjected to rather high-momentary flow rates. If the flow rate exceeds the rating of the dryer, then wet air will pass and be introduced to the distribution system. Most users select a dryer that has a flow rate that matches the continuous flow rate of the

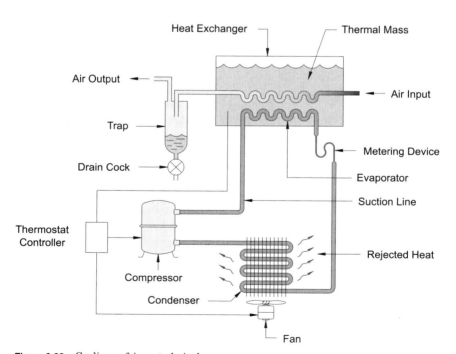

Figure 6-22 Cycling refrigerated air dryer.

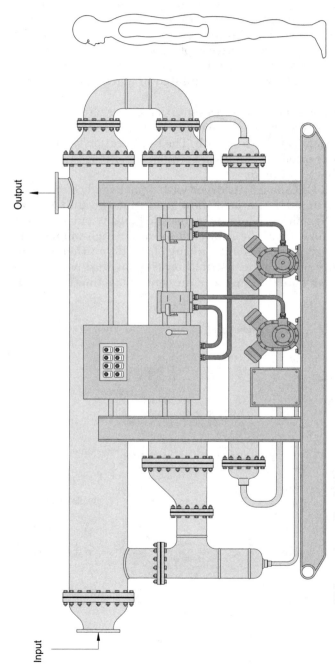

Input

Output

Figure 6-23 High-capacity cycling dryer.

Air Output

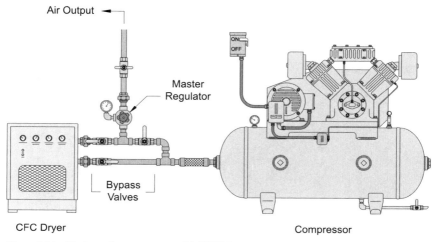

Master Regulator

Bypass Valves

CFC Dryer

Compressor

Figure 6-24 Packaged compressor with/CFC dryer.

compressor pump. However, this can be a little deceiving, the receiver is a storage device and can easily deliver five times the pump rate during peak load periods. If these peak load periods are infrequent they generally do not represent a problem. On the other hand, if peak load periods are frequent then the dryer must be sized to match this demand. Figure 6-24 shows a typical dryer installation on a packaged two-stage reciprocating compressor. The bypass valves allow the dryer to be taken off line for service without disturbing air delivery to the plant. Note that the master regular is after the dryer. Placing the regulator after the dryer will help to mitigate surge effects.

When installing a CFC-based air dryer within a screw compression system, the flow rate of the dryer should match the maximum rate of the compressor pump. Normally the dryer is installed between the aftercooler and the receiver. Since the receiver has dry air, high-momentary flow rates can be supplied without affecting the dryer function. If the flow rate momentarily exceeds the rating of the dryer, it will have no effect on the quality of the delivered air. Figure 6-25 shows a typical screw compression system with a CFC-based dryer. Note that an aftercooler is used between the compressor and the dryer. The aftercooler can usually be deleted if a high-temperature dryer is selected.

Elliott Cycle Refrigerated Dryers

Elliott cycle refrigerated compressed air dryers operate in generally the same way that any other refrigerated dryer operates. The compressed air is cooled in a controlled environment, condensing water is trapped

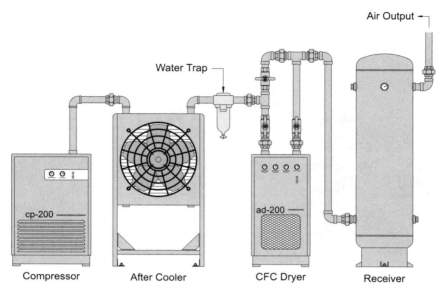

Figure 6-25   Screw compressor with/CFC dryer.

and then periodically drained off. The principal difference is the simplicity of the refrigeration cycle and the durability of the equipment.

The Elliott refrigeration cycle is based on fluid evaporation. Figure 6-26 shows a schematic representation of the cycle. Compressed gas is injected through the injector gas nozzle, which, in turn, drives a two-stage Venturi. The first-stage Venturi pulls a negative pressure within the fluid chamber. A volatile fluid is allowed to bleed into the fluid chamber. The amount of fluid that bleeds into the chamber is controlled with a needle valve (fluid control). As the fluid enters that chamber, it is

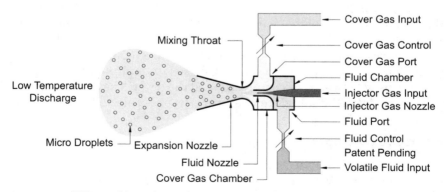

Figure 6-26   Elliott refrigeration cycle.

atomized and forms micro-droplets. This atomized fluid is then forced out through the fluid nozzle.

The second-stage Venturi pulls a negative pressure within the cover gas chamber. A dry cover gas is introduced into the chamber. The amount of cover gas that is allowed to flow into the chamber is controlled by a second needle valve (cover gas control). The cover gas and the atomized water are forced together as they pass through the mixing throat. As the mixture enters the expansion nozzle, the atomized fluid realizes a high rate of evaporation and, consequently, a low discharge temperature is generated.

The cover gas can be any gas with a high affinity for the fluid. Dry nitrogen is an excellent choice for many fluids such as water, acetone, alcohol, MEK, and gasoline. For lower-performance applications, tap water, compressed air, and ambient air can be used for the fluid, injector gas, and cover gas. Ambient air, as a cover gas, will benefit greatly by predrying through a desiccant filter.

Refrigerated compressed air drying is a relatively low-performance application, therefore, water, compressed air and ambient air are used for the fluid, injector gas, and cover gas. These provide an excellent, low hazard, and environmentally friendly solution for this requirement.

In practice, the refrigerator body (Fig. 6-27) is a piece of molded plastic with a plastic water vapor nozzle pressed into the core. The air nozzle is a brass piece that is pressed into the core of the water vapor nozzle. The space formed between the OD of the air nozzle and the core of the water vapor nozzle is the water vapor chamber. The space formed between the OD of the water vapor nozzle and the core of the body is the cover gas chamber. Each chamber and the air nozzle are ported

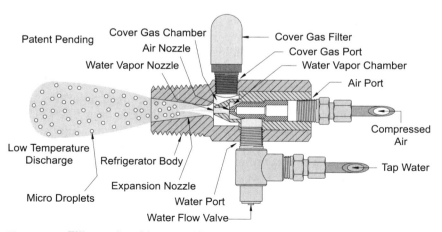

Figure 6-27 Elliott cycle refrigerator. (*Courtesy of Air Options Inc.*)

with a $1/8$-in. National Pipe Thread (NPT) thread. The cover gas port mounts a 100-μm filter and the water port mounts a brass-needle valve with a $1/4$-in. compression fitting. It should be noted that this particular arrangement does not require a needle valve to control the cover gas. The nose of the body carries a $1/2$-in. male NPT thread, which allows it to be screwed directly into the shell of a heat exchanger.

This particular design is specifically intended for compressed air drying applications. The amount of water that is introduced into the unit is more than it can be completely evaporated. The reason for this is that the micro droplets that do not evaporate, normalize to same temperature as the gas within the discharge. These cold micro droplets then come in contact with the tubes within the heat exchanger and allow for a high-efficiency thermal transfer.

The construction of these dryers is extremely rugged (Fig. 6-28). The body of the unit is all-welded Sch 40 piping components and the exchanger tubes are stainless steel. The bottom of the heat exchanger carries a trap with an integral coalescing filter. All ports on the welded assembly are male NPT threads. The refrigerator(s) mount into ports at the top of the heat exchanger. Units in the 6- to 120-SCFM range mount directly to the compressors they service. Dryers 150 SCFM and larger are floor mounted. The 6- to 120-SCFM units have a control system that consists of two solenoid valves, a water valve, and an air valve. The valves are controlled by the motor controller on the compressor. When the compressor motor is on, then the dryer is on. When the compressor motor is off, then the dryer is off. Dryers 150 SCFM and larger are controlled manually.

Because the coolant is a vapor cloud, the cross section of the heat exchanger must be limited to a section so that the vapor discharge can effectively penetrate. This cross section limits the construction of the heat exchangers to units that have approximately 500 SCFM capacity. To manufacture larger capacity dryers, 500 SCFM heat exchangers are welded together on common manifolds as shown in Fig. 6-29. Elliott cycle dryers are manufactured in this fashion up to 3000 SCFM (six 500 SCFM heat exchangers). For dryers over 3000 SCFM, a bolt-together manifold is manufactured and smaller units are ganged together. As an example, a 10,000 SCFM dryer is assembled utilizing four 2500 SCFM units on common manifolds and with common controls. Figure 6-29 shows a 2000 SCFM dryer assembled using four 500 SCFM heat exchangers on a common set of manifolds.

Elliott cycle dryers are a type of refrigerated dryer that does not use CFCs, but has the same role as CFC dryers without all their drawbacks. They are specifically designed to be extremely compact and rugged. Figure 6-30 shows a schematic representation of a packaged reciprocating air compressor with an Elliott cycle dryer. The dryer is

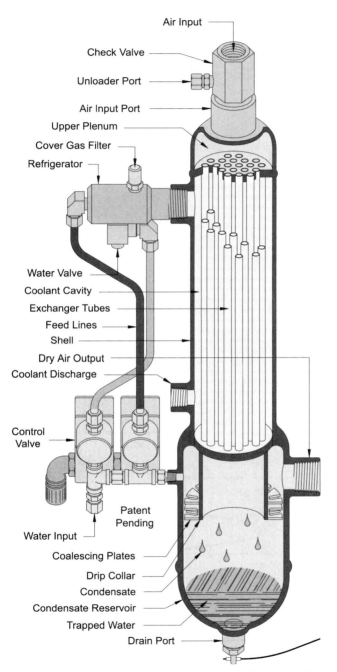

Air Input

Check Valve

Unloader Port

Air Input Port

Upper Plenum

Cover Gas Filter

Refrigerator

Water Valve

Coolant Cavity

Exchanger Tubes

Feed Lines

Shell

Dry Air Output

Coolant Discharge

Control Valve

Water Input

Patent Pending

Coalescing Plates

Drip Collar

Condensate

Condensate Reservoir

Trapped Water

Drain Port

Figure 6-28 Elliott cycle dryer sectional view. (*Courtesy of Air Options Inc.*)

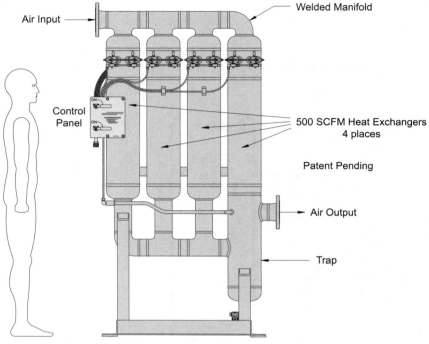

Figure 6-29   2000 SCFM Elliott cycle dryer. (*Courtesy of Air Options Inc.*)

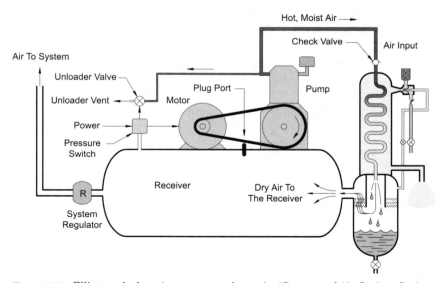

Figure 6-30   Elliott cycle dryer/compressor schematic. (*Courtesy of Air Options Inc.*)

installed between the output of the pump and the input of the receiver. The un-loader, pressure switch, and motor work in the same fashion as they did before installation. The in-tank check valve is removed and plugged. The output of the dryer is mounted to the end port of the receiver. The output of the pump is plumbed into the input of the dryer. If the unloader line was connected to the in-tank check valve, then it should be reconnected to the $\frac{1}{8}$-in. NPT port on the top of the dryer. If the unloader line is connected to the head of the pump, then it should be left in place. The discharge should be plumbed to a suitable drain and a tap water source connected to the dryer.

Standard Installations

Figure 6-31 shows a typical installation on a vertical pattern reciprocating compressor. Note that the unloader line is connected to the check valve on the top of the dryer.

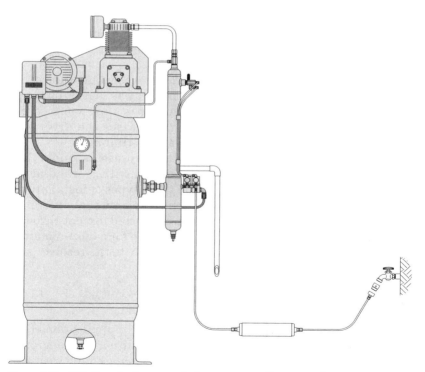

Figure 6-31   Elliott cycle dryer, vertical compressor. (*Courtesy of Air Options Inc.*)

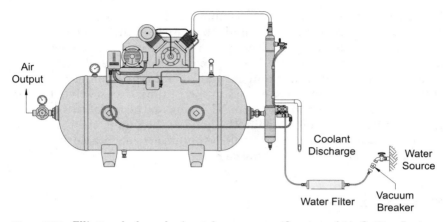

Figure 6-32   Elliott cycle dryer, horizontal compressor. (*Courtesy of Air Options Inc.*)

Figure 6-32 shows a typical installation on a horizontal pattern reciprocating compressor. Note that the unloader line is connected to the head of the pump.

Screw compressors have traditionally been reserved for systems over 30 hp. In recent years, however, these compressors have started to encroach on the smaller market. Screw compressors in the 10- to 30-hp range are becoming increasingly more prevalent.

Elliott cycle dryers in the 6- to 120-SCFM range are installed and operate on screw compressors in essentially the same way as they do on reciprocating compressors. The only real exception is that some screw compressors are continuous-run units. In these cases the dryer will operate continuously, as long as the compressor is running.

Screw and, less commonly, reciprocating compressors under 120 SCFM may be installed as a base-mounted unit with a stand-alone receiver. Mounting an Elliott cycle dryer on these types of installations is principally the same as a packaged system. The only real difference is that the compressor and receiver are not connected, except through the piping. Figure 6-33 shows a typical installation for a base-mounted compression system. The dryer is mounted directly to the receiver and the air line is plumbed to the input of the dryer.

Duplex Installations

Duplex compressors are rather common systems for applications that have high peak load demands or systems that require a high degree of redundancy. These installations may be either a single unit with two pump/motor assemblies, as shown in Fig. 6-34, or two completely

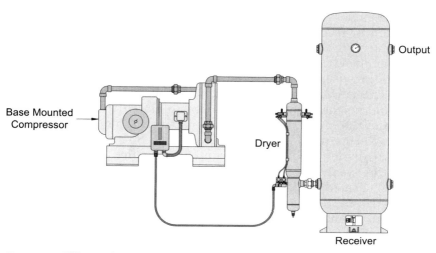

Figure 6-33   Elliott cycle dryer, base mounted compressor. (*Courtesy of Air Options Inc.*)

separate compressors with common controls. In either case, the installation of an Elliott cycle dryer is the same.

The two pump/motor assemblies must be able to operate independently and together. Additionally, each assembly must be serviceable without affecting the operation of the other. To accomplish this, the dryer must be equipped with a simple input manifold carrying

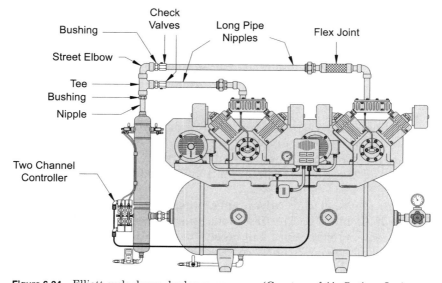

Figure 6-34   Elliott cycle dryer, duplex compressor. (*Courtesy of Air Options Inc.*)

two check valves. The check valves should be "compressor rated." The manufacturer recommends the use of ordinary in-tank check valves for this purpose. The input manifold is constructed using a 300-lb tee and street elbow that are one size up from the dryer size. Forged steel bushings are used to match the manifold to the dryer and check valves. The manifold is completed with the addition of the check valves and a schedule 80 nipple, sized to match the output of the pumps and the input of the dryer, respectively. The pumps are connected to the manifold with two long, schedule 80 nipples. The longer of the two nipples should be interrupted with a flex joint to allow for differential expansion.

Control of the dryer is accomplished by replacing the standard controller with an optional two-channel controller. The dryer size should be matched to the maximum SCFM of the system. If the system is used as a peak demand system, then the dryer size should be equal to the output of the pumps running simultaneously. That is, if the system has two 10-hp pumps, then an 80-SCFM dryer should be selected. If the system is intended for redundancy only, then the dryer should be sized to match the larger of the two pumps.

Applications Greater Than 120 SCFM

Elliott cycle dryers are generally stand-alone units for systems that produce air flows greater than 120 SCFM. The dryer is installed between the compressor and the receiver. There is no requirement for an aftercooler in these applications. If a dryer is being fitted to a peak load duplex system, then it should be sized for the combined output of both compressors. If the system is intended for redundancy only, then the dryer should be sized to match the larger of the two compressors. Figure 6-35 shows a typical installation on a redundant screw system.

Instrumentation Air Systems

For instrument air systems, an Elliott cycle dryer should be applied between the compressor and the "wet" or utility air receiver. If there is no utility air requirement, then the utility air receiver is not required and the Elliott cycle dryer should feed the desiccant dryer directly.

There are three distinct advantages of using an Elliott cycle dryer in this type of installation. The first is that the temperature of the air that the desiccant dryer receives will be less than 55°F and, therefore, the water load will be significantly reduced. This means that the desiccant charge in the "drying" tower will not saturate halfway through the regeneration cycle and consequently contaminate the system. The

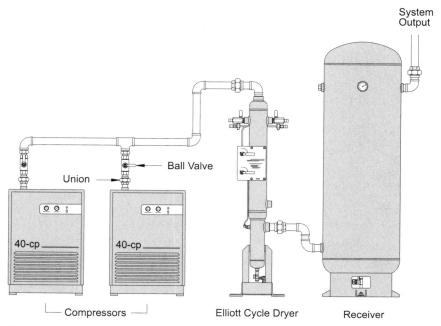

Figure 6-35   Elliott cycle dryer, duplex system. (*Courtesy of Air Options Inc.*)

second benefit is that oil contamination is principally eliminated, which translates to significantly longer desiccant life. The third advantage is that Elliott cycle dryers are extremely rugged and will provide a long term or permanent solution for first-stage drying. Figure 6-36 shows how an Elliott cycle dryer would be applied to an instrument air system with a utility air output.

Saltwater Prepared Dryers

Elliott cycle dryers are frequently used in offshore applications and operate with saltwater as their coolant. This is not recommended for the standard dryers; however, manufacturers do offer a special saltwater construction that makes these dryers compatible with marine applications. To prepare a dryer for this duty, all the components of heat exchanger are protected with corrosion-resistant materials. The trap and head assemblies remain of carbon steel and their internals are treated with a corrosion resistant coating. All fittings and external tubing are replaced with corrosion-resistant components and the units have no moving parts. Figure 6-37 shows a sectional view of all the corrosion-protected components used in the assembly.

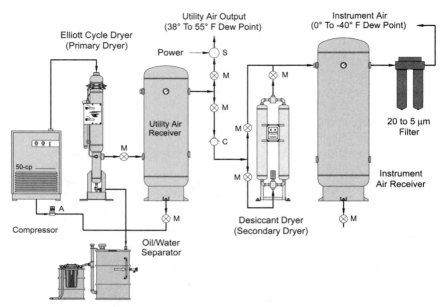

Figure 6-36 Elliott cycle dryer, instrument air system. (*Courtesy of Air Options Inc.*)

Ice Bath Dryers

A variation of the refrigerated dryer is the ice bath dryer. This type of dryer is just as effective as its CFC counterpart except for one significant drawback. These dryers require a fresh charge of ice every few hours of operation. This is impractical for continuous use compression systems. However, for systems that are used intermittently, these dryers can offer an excellent, low-cost solution for drying compressed air. They are exceptionally effective for farming operations, home garage, low-use maintenance facilities, school and club shops, field maintenance, contractors, and the like.

Their operation and construction is extremely simple. They consist of a coil that is immersed in water/crushed ice bath. The coil has a water trap at the bottom to collect condensate. The compressed air is fed through the coil and is cooled to the temperature of the ice bath. The water vapor in the air condenses onto the inside of the coil and runs down, along with the airflow, into the trap. The dried air is taken off the top of the trap. Figure 6-38 shows a schematic representation of an ice bath dryer.

An ice bath dryer is a simple, compact piece of equipment. They are normally built into some sort of standard insulated cooler. They can be charged with 10 lb of ice in the morning and will support three or four

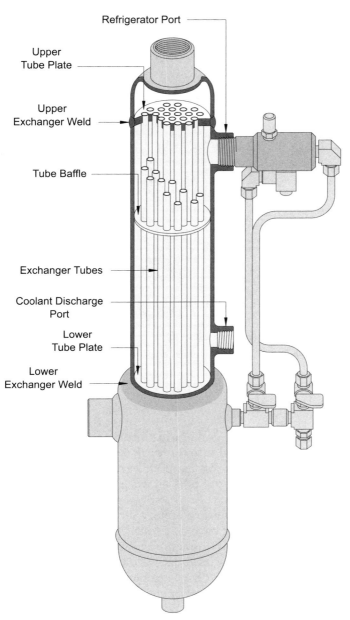

Figure 6-37 Salt water prepared dryer. (*Courtesy of Air Options Inc.*)

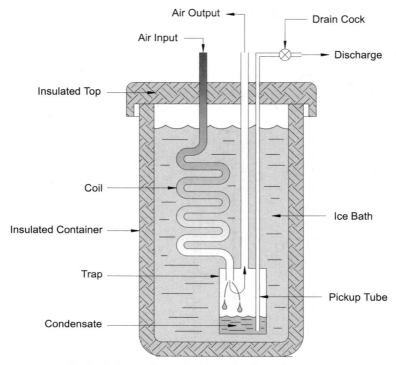

Figure 6-38 Ice bath dryer schematic. (*Courtesy of Air Options Inc.*)

nail guns for the entire day. At the end of the shift the water is simply dumped out and the dryer is loaded into the work truck along with the compressor. Setting up the dryer with a manifold carrying several quick disconnects is very convenient. Figure 6-39 shows a sectional view of a commercial 20 SCFM ice bath dryer.

Setting up and using ice bath dryers is very straightforward. The dryer is charged with an ice-water bath and placed in a location adjacent to the compressor. The output of the compressor is connected to the input of the dryer. The hose that is used between the compressor and dryer should be at least $^3/_8$-in. ID and kept as short as possible, generally less then 10 ft. The application hose is connected to the output of the compressor and the system is ready for use. The dryer's drain should be vented as often as necessary and the ice charge should be checked every two hours. If the ice charge is significantly reduced then the dryer should be recharged. Figure 6-40 shows a commercial ice bath dryer set up with a small single-stage compressor.

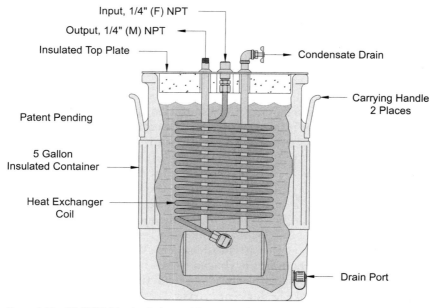

Figure 6-39 20 SCFM ice bath dryer. (*Courtesy of Air Options Inc.*)

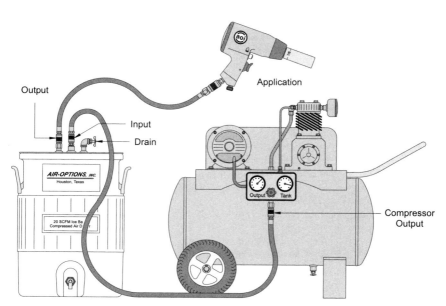

Figure 6-40 Ice bath dryer and compressor. (*Courtesy of Air Options Inc.*)

Desiccant Dryers

For applications that require a higher degree of dry air, a second-stage desiccant dryer is the normal solution. These dryers have a charge of media that traps water vapor. Because desiccant dryers are "getter" devices, their overall capacity is limited. For situations that require very low flow, replaceable element units may be deployed.

Figure 6-41 shows a schematic representation of a desiccant dryer. Air is directed through the input and down to the bottom of the charge. The air is forced to turn 180° and flow up through the desiccant charge. When the air reaches the top of the charge, it is allowed to exit the output. As the air travels through the desiccant charge, airborne water vapor is absorbed. Starting at the bottom, the charge becomes saturated, and with use the saturation moves toward the top. Most desiccant dryers have a sight glass at the top of the housing. As the desiccant saturates, it will turn pink. When a pink color is seen in the sight glass it is time to replace or regenerate the desiccant. Usually, desiccants can be regenerated by spreading the charge over a baking sheet and heating it to an elevated temperature for a period of time. Figure 6-42 shows an ordinary single canister desiccant dryer.

Desiccant dryers are the most misapplied dryer made. These dryers are normally placed into service as a primary dryer, and in this role the water load is entirely too large. Desiccant dryers are intended for

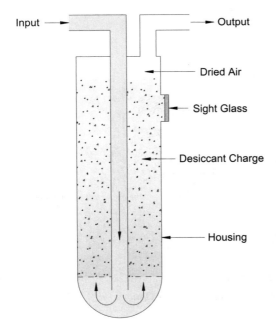

Input ⟶

Output ⟶

Dried Air

Sight Glass

Desiccant Charge

Housing

Figure 6-41 Desiccant dryer schematic.

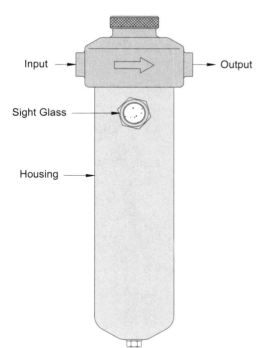

Input —

Output

Sight Glass —

Housing —

Figure 6-42   Canister dryer.

applications that require very clean air such as fine finishes, semiconductor manufacturing, and instrument air.

A desiccant dryer normally requires a first-stage dryer such as a
refrigerated dryer. A significant drawback to these types of dryers is that
the desiccant is very sensitive to oil contamination. Oil can ruin the desiccant, so it is imperative that a first-stage dryer removes all the oil from
the air stream before it is introduced to the desiccant charge. If oil contamination occurs, the only solution is to replace the desiccant charge.

Another consideration is that the input temperature to a desiccant
dryer must be kept rather low, preferably less than 70°F. Feeding high-
temperature air to a desiccant dryer will greatly diminish its performance. High-temperature air carries a substantially greater water load
than cooler air. As an example, air at 100 percent relative humidity and
100°F has twice as much water content as air at 100 percent relative
humidity and 78°F.

Single-canister desiccant dryers should be used as a second-stage dryer
only. These dryers should only be fed air that has been previously dried
through some type of refrigerated dryer. If unprocessed air is introduced
to a canister desiccant dryer, it will quickly saturate and its function will

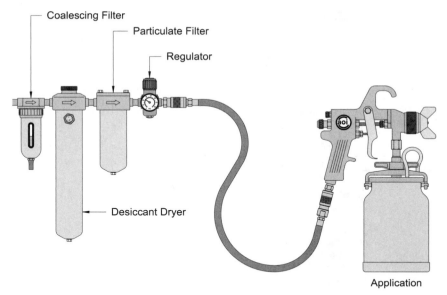

Figure 6-43   Desiccant dryer at application site.

be negated. Figure 6-43 shows a typical arrangement when using a single-canister desiccant dryer. A coalescing filter is placed before the dryer to trap any liquid that may be contaminating the input air. The output of the dryer should feed a 5- to 20-μm particulate filter. The filter is necessary because desiccants have a tendency to "dust." In particular, low-cost desiccants can introduce fine particles into the airstream. The output of the filter should carry the application regulator. The arrangement illustrated is ideal for automotive and fine-finish applications.

For high-flow applications, a regenerative twin tower dryer is preferred (Fig. 6-44). This type of dryer has two different desiccant charges and while one is drying air, the other is being regenerated. Regeneration of desiccants is accomplished by heating the desiccant—forcing the trapped water to vaporize and venting it away. Twin-tower units have two diverter valves, which control the flow of air during this regeneration cycle. The tower that is being regenerated is heated and dry air from the output of the opposite tower is flowed through in order to carry off the water vapor to a vent. After a period of time, the cycle flips and the towers trade duties.

Like any desiccant dryer, twin tower dryers require relatively low-input temperatures. If the input temperature is too high, then the dryer cannot regenerate as quickly as it absorbs water and the system is rendered ineffective. Additionally, the higher the air temperature, the higher the oil load, which means a certain death for desiccants.

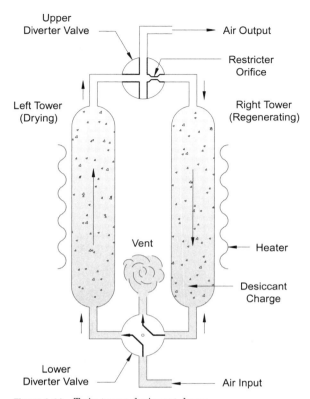

Figure 6-44 Twin tower desiccant dryer.

When specifying a twin tower desiccant dryer it is important to consider its air and energy consumption. It is commonly believed that desiccants are fairly energy efficient; this is not the case. Generally 25 SCFM of airflow and 3 kW of electric heat are required to regenerate 100 lb of desiccant. Since desiccant dryers regenerate continuously, this means their annual operating costs can be significant. As an example, a 250-SCFM twin tower desiccant dryer carrying two 200 lb charges of desiccant requires 2 hp of compressed air and 6 kW of heat on a continuous basis. That's equal to 70,080 kWh, or roughly $2800.00 per year of operating expense!

Twin tower desiccant dryers are generally placed in a systems role. They can commonly be found in or adjacent to the compressor room. Figure 6-45 shows a typical twin tower installation. The dryer must have some sort of first-stage dryer such as a refrigerated dryer or high-efficiency aftercooler. Twin tower desiccant dryers should always be backed up with a receiver. These dryers do not perform well during surge events.

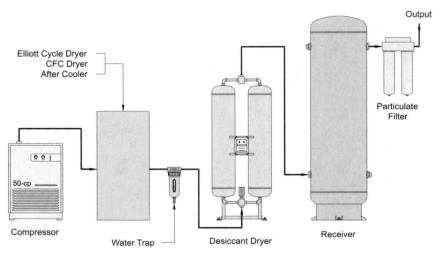

Figure 6-45   Low dew point system.

Deliquescent Dryers

Deliquescent, or what are commonly referred to as single tower dryers, occupy a small niche in the dryer industry. These dryers are little more than a large pressure vessel filled with salt. Salt has a high affinity for water and, therefore, is used for some drying applications. Figure 6-46 shows a schematic representation of a deliquescent dryer. Air is introduced into the bottom of the vessel and travels up through the salt charge. As the water vapor in the air comes in contact with the salt, the salt melts and drips down into the bottom of the tank. Deliquescent dryers will normally provide a 20°F shift in the dew point. Their primary drawback is that their media (salt) is highly corrosive. Other drawbacks are that they are the largest of the dryer types and the salt charge must be refilled on a periodic basis. Additionally, these dryers do not work well in warm climates. The salt has a tendency to melt at an accelerated rate when exposed to high input or ambient temperatures. Deliquescent dryers are normally deployed in conjunction with some type of first-stage dryer, such as a refrigerated dryer or a water-cooled aftercooler. Figure 6-47 shows a typical deliquescent dryer. These dryers are very ordinary in appearance and to the untrained observer they can easily be mistaken for an oversized receiver.

Membrane Dryers

One type of specialty dryer on the market is the membrane dryer. These dryers are typically used for point-of-use applications. There are some

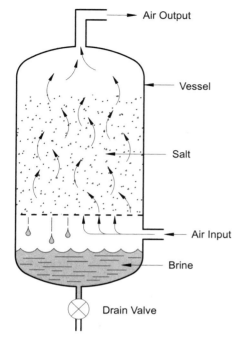

Figure 6-46 Deliquescent dryer schematic.

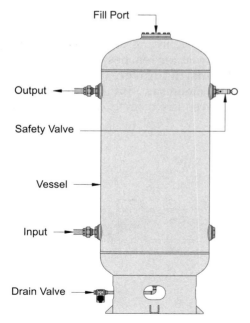

Figure 6-47 Deliquescent dryer.

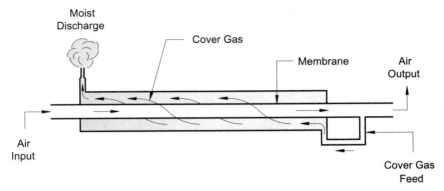

Figure 6-48   Membrane dryer schematic.

large membrane dryers manufactured, but they are rather uncommon. Membrane dryers are most commonly found in laboratory applications.

Figure 6-48 shows a schematic representation of a membrane dryer. Air is flowed over a membrane, which has a high affinity for water vapor. As the water builds up in the membrane, it migrates through to the other side. A small amount of dry air, referred to as the cover gas, is drawn from the output and flowed over the opposite side of the membrane. The water in the membrane evaporates into the cover gas and is carried out through the discharge. Figure 6-49 shows a typical commercial membrane dryer. Note the coalescing filter on the input.

The most significant disadvantage of the membrane dryer is surge situations. If the dryer is surged, then wet air reaches the output end and, therefore, the cover gas is wet. This, in turn, greatly reduces the evaporation rate of the water in the membrane into the cover gas. After a surge situation, these dryers require a great deal of time to recover.

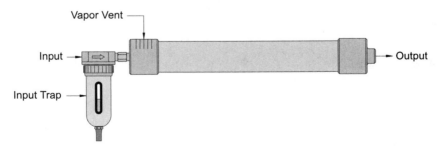

Figure 6-49   Commercial Membrane Dryer.

Absorption Dryers

An interesting type of dryer, which is worth mentioning for discussion purposes only, is the absorption dryer. These are very uncommon dryers and are usually found only in extremely large applications such as bulk natural gas drying.

The absorption dryer (Fig. 6-50) is simple in operation. A large charge of fluid, which has a high affinity for water, usually elethlene glycol, is contained in a vessel. The air to be dried is bubbled up through the fluid, which, in turn, absorbs any water vapor in the air. The dry air output is at the top of the vessel. The water-saturated fluid is allowed to pass into a degassing sump where it is heated and the water content boils off. The cleaned fluid is then pumped back into the vessel.

Dryer Performance, Types, and Applications

The indicator that is used to describe the dryness of compressed air is dew point. The dew point is the temperature at which water will start to condense out of air. Suppose that we expose air to, say, a piece of glass, which has a decreasing temperature range starting at 110°F and finishing at 38°F. As we watch the glass, there will come a time that water starts to condense on its surface. If the temperature of the glass is 70°F at the time that condensation starts, then the dew point of the air is 70°F. Dryer performance is always stated in dew point, so it is very important to understand this scale. The dew point scale shown in Fig. 6-51 references common dryer types against common compressed air applications.

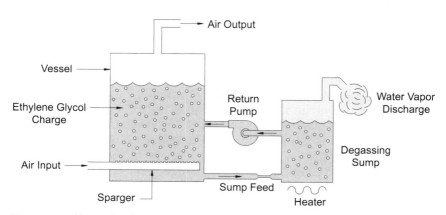

Figure 6-50 Absorption dryer.

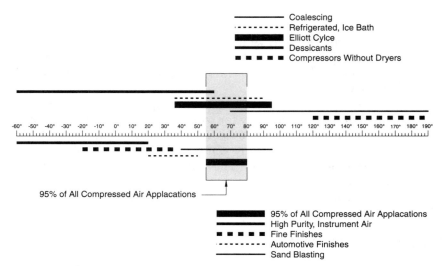

Figure 6-51   Dew point scale.

Oil/Water Separators

The water that is drained from any compression system is considered hazardous waste and must be handled properly. Simply blasting the condensate out into the back driveway has become unacceptable behavior. State and federal regulations are rather specific about this form of dumping and heavy fines and cleanup cost can be levied against any company that continues this practice. To further complicate this situation, the cost associated with handling the shear quantity of condensate that any compression system generates is prohibitive.

The actual contaminates in the discharge are really very small and can be separated by utilizing a commercial oil/water separator. These systems are inexpensive, extremely simple to operate, require very little maintenance, and sufficiently clean the compressor discharge so that it may be dumped into the storm sewer.

The separator works in two stages (Fig. 6-52). The first is a large reservoir that allows the oil within the condensate to separate via gravity. Condensate is introduced at the top of the reservoir through a diffuser. The diffuser is intended to slow the velocity of the input so that it does not remix the oil/water that is already in the reservoir. The reservoir is sized to allow the oil/water enough resident time so that the oil can separate by floating to the top. The bottom of the reservoir is clear water. The clear water exits through the bottom, via the standpipe

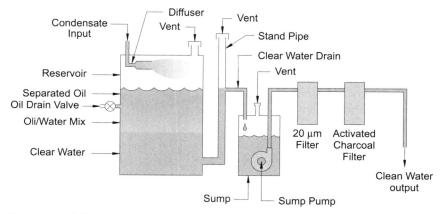

Figure 6-52   Oil/water separator. (*Courtesy of Air Options Inc.*)

and clear water drain. The top of the standpipe is vented to prevent siphoning action. A sight glass is incorporated so that the amount of separated oil may be monitored. When enough oil is separated, it may be drained off through the oil drain valve and sent to an appropriate oil-recycling center.

The clear water that is discharged from the separator has trace amounts of contaminates and must go through a filtration process. The second stage is a sump with an integral sump pump. As the clear water drains, the sump slowly fills to a level that will activate the pump. The pump forces the clear water through a 20-μm filter and then an activated charcoal filter. The 20-μm filter removes any particulate contamination and the activated charcoal filter removes any trace chemical contamination. This type of separation produces excellent water quality and can be installed on virtually any size compression system.

Commercially available oil/water separator systems are simple and reliable. They typically consist of a larger separation reservoir and a smaller sump. The sump contains a pump which feeds two filter housings mounted on the outside. These pumps normally operate on 115 VAC, but they are also available in 220 VAC. The filter elements are readily available from a variety of sources as well as from the manufacturer of the separator. When sized and set up properly, these separators only require maintenance about once a year. This maintenance amounts to draining the separated oil and replacing the filter elements. Figure 6-53 shows a commercial oil/water separator.

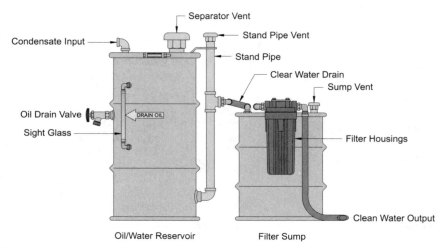

Figure 6-53 Commercial oil/water separator system. (*Courtesy of Air Options Inc.*)

Questions*

1. Name two common traps.
    (A) Line   (B) Cyclone   (C) Mouse   (D) Marriage

2. Name two types of coalescing filters.
    (A) Porous media   (B) Desiccant   (C) Plate   (D) Micro grid

3. What are the two most common aftercoolers?
    (A) Water   (B) Oil   (C) Air   (D) Injector

4. What is the most common "effective" dryer type?
    (A) Absorption   (B) Membrane   (C) Refrigerated   (D) Desiccant

5. Between what two components is an Elliott cycle dryer installed?
    (A) Tank and regulator   (B) Piping and tools   (C) Pump and receiver
    (D) Motor and plug

6. What is the best dryer type for contractors?
    (A) Desiccant   (B) Ice bath   (C) Membrane   (D) Absorption

7. What type of dryer is preferred for high purity applications?
    (A) Absorption   (B) Membrane   (C) Refrigerated   (D) Desiccant

8. What type of media is used in a deliquescent dryer?
    (A) Desiccant   (B) Cotton   (C) Salt   (D) Sand

9. If water starts to condense at 85°F, what is the dew point of the air?
    (A) 65°F   (B) 96°F   (C) 85°F   (D) 81°F

10. What is the dew point range for most compressed air applications?
    (A) 70°F–80°F   (B) 38°F to 90°F   (C) –40°F–10°F   (D) 0°F–38°F

---

*Circle all that apply.

Support Components

Any compressed air system requires a wide variety of bits and pieces to make it function properly. This is true of a simple home-shop compressor and large industrial systems alike. Oftentimes, a well-rounded knowledge in this area of compressed air provides the broadest-reaching benefit to those personnel tasked with the maintenance and operation of a compressed air system. Those personnel are well advised to study this chapter with great care. The components outlined here act as the glue that binds the compression system together and can make or break any installation. This chapter provides a brief outline of some of the most common support components, from basic pipe fittings to complex control devices. Figure 7-1 shows just a few of the components that we will

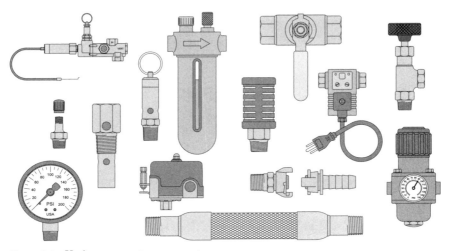

Figure 7-1   Various support components.

be reviewing in this chapter. After familiarizing yourself with these items, you might decide that additions should be included in your current system or in the system that you're designing. It might also be a good idea to stock certain parts for repair and replacement. Surprisingly enough, most of the components in this chapter are very inexpensive and oftentimes it is cheaper to stock a critical component then it is to endure the downtime while you go get a replacement part.

Filters

Within any compressed air system, there exists a requirement for filtration. The compressor will generate a certain amount of particulate contamination; however, the real problem area is the internals of the distribution system. The inside of your receiver and pipes are constantly rusting, creating particles, which flake off and move along with the flow of the compressed air. Ultimately, this particulate contamination arrives at the application where it can do considerable damage. To counter this inevitable problem, particulate filters should be applied at each application and at the output of any device that may generate contamination. Figure 7-2 shows a particulate filter housing that uses standard replicable cartridge filters and a pleated paper filter element. These elements are very inexpensive and are available in a wide range of capacities and micron sizes.

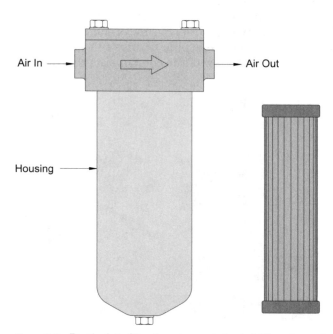

Figure 7-2 Particulate filter housing and pleated filter element.

Pressure Regulators

Most compressed air systems require at least one regulator, and in many situations they require a large number of these devices. A pressure regulator is a control device that takes a high input pressure and lowers it to a usable output pressure. The most important regulator in the system is the master regulator, which is responsible for controlling the primary pressure in the entire distribution system. These regulators are usually placed on the output of the primary receiver. Generally the output of a master regulator should be set at 90 psi. It should also be noted that the regulator's flow rate must be ample to deliver full pressure even during peak surge conditions. A good rule of thumb is to select a master regulator that is capable of delivering a flow rate that is at least five times the flow rate of the compressor. If you have a 10-hp compressor with a rating of 38 SCFM, then a regulator should be selected that has a flow capacity of 380 SCFM. Figure 7-3 shows a typical master regulator. Smaller regulators are applied at the different application sites as necessary.

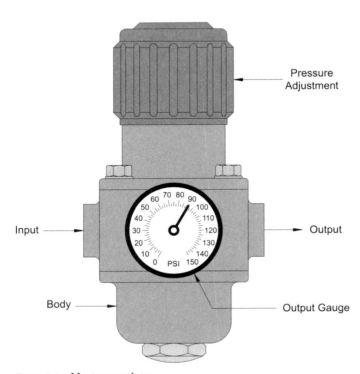

Figure 7-3   Master regulator.

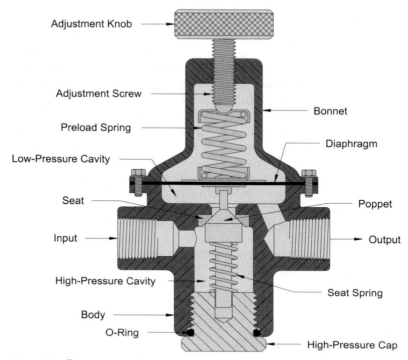

Adjustment Knob

Adjustment Screw

Bonnet

Preload Spring

Diaphragm

Low-Pressure Cavity

Seat

Poppet

Input

Output

High-Pressure Cavity

Seat Spring

Body

O-Ring

High-Pressure Cap

Figure 7-4   Pressure regulator.

Figure 7-4 shows a sectional view of a pressure regulator. High-pressure air is introduced to the high-pressure cavity via the input port. The poppet and seat provide a valve that is used to control the flow of air into the low-pressure cavity. The top of the low-pressure cavity is in the form of a diaphragm. The diaphragm is backed with a preload spring, which can be adjusted via an adjustment screw and knob. When a preload is applied, the diaphragm pushes the poppet down and allows air to flow from the high-pressure cavity to the low-pressure cavity. When the output pressure becomes high enough, it applies a force to the diaphragm that counters the preload spring and stops or restricts the flow between the cavities. Pressure regulators generally have two additional gauge ports that allow the operator to monitor the input and output pressures.

Lubricators

Air tools, especially those that use air motors, will benefit from proper lubrication. This is an area of compressed air that is usually neglected, although it is responsible for causing millions of dollars of damage to air

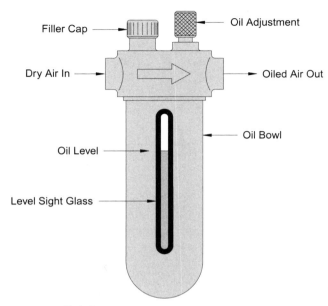

Figure 7-5 Lubricator.

tools every year. If an air tool is lubricated properly, it will last a life-time. On the flip side, if the lubrication of an air tool is neglected, the tool will have a particularly short life. Most manufacturers of pneumatic equipment provide lubrication recommendations with their products. Pneumatic devices will provide exceptional performance by closely following these recommendations. Figure 7-5 shows a typical in-line lubricator. These units generally have a reservoir, sight glass, filler cap, and flow control. They can be permanently placed at air ports and any tool that is connected to the port will receive proper lubrication. It should be noted that if there are applications that require oil-free air, a second non-oiled port should be provided. Similarly, hoses that are used for oiled applications should not be used with non-oiled applications.

Filter, Regulators, and Lubricators Guide

For most applications, it is necessary to provide some sort of secondary air processing. This varies from application to application. Figure 7-6 shows some common uses for compressed air and how the air should be processed. This particular arena is the most neglected of the compressed air system. It is almost always an afterthought and treated more as an annoyance than a legitimate segment of the system. Installing carefully thought out secondary processing will have a profound effect on the life and performance of your air tools, equipment, and processes.

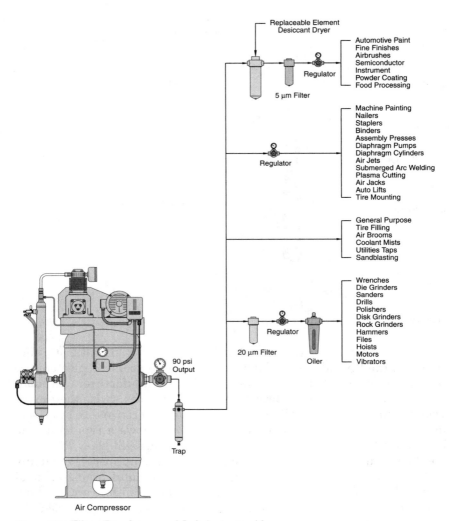

Figure 7-6 Filter, Regulators and Lubricators guide.

Safety Relief Valves

Safety relief valves are a critical safety device that *must* be used on all compression systems. These valves prevent catastrophic failure of the essential pressure equipment and piping. All receivers *must* be equipped with a safety relief valve that is rated to open at a pressure no higher than the maximum rating on the receiver's name plate. The valves *must* be equipped with a manual actuator. Compression pumps should be equipped with safety relief valves to prevent costly damage to the

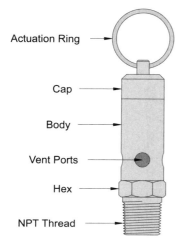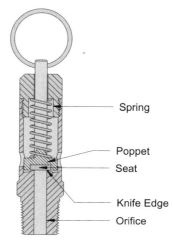

Figure 7-7 Safety relief valve.

internals of the pump in the event of a control failure. Safety valves should be inspected on a regular basis and verified in accordance with the manufacturer's recommendations. Valves that are found to be leaking, in poor condition, excessively dirty, missing the manual actuator, or showing any damage should be replaced immediately. The afflicted valve should be discarded or sent back to the manufacturer for overhaul and requalification. Safety valves are generally not adjustable and are delivered by the manufacturer at a preset pop-off pressure. Figure 7-7 shows a typical safety relief valve. The OD of the poppet is considerably larger than the orifice size. This oversize design allows the valve to open at a high pressure and remain open until the pressure has dropped to a considerably lower pressure, usually about half of the valve's pop-off pressure. To determine the close pressure of a safety valve, simply turn off your compressor and manually actuate the valve. The pressure of the system when the valve stops flowing is the close pressure.

Pressure Relief Valves

A similar valve like the safety valve is the pressure relief valve. These valves are intended for a different role than their safety counterparts. A pressure relief valve is intended to vent excess pressure from the system. These valves are, in a sense, a type of pressure regulator. When the pressure exceeds a preset level, the relief valve opens and bleeds down the system. These valves normally have an external adjustment so that the cracking pressure may be tuned to match any given application. The poppets of these valves are generally designed with a tapered

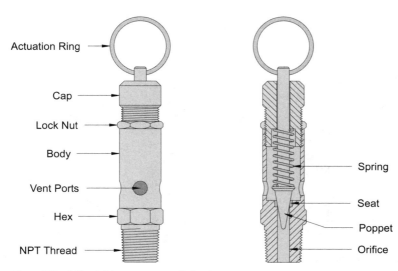

Figure 7-8 Adjustable pressure relief valve.

plug so that they only bleed the excess pressure and reclose at their set point. Pressure relief valve adjustments can be found in a variety of forms, from a threaded cap and lock nut to a micrometer barrel. These valves usually carry some type of manual actuator. Figure 7-8 shows a typical pressure relief valve.

> Caution: Many pressure relief valves have a similar appearance to their safety counterparts. Under no circumstances should a pressure relief valve be used as a safety valve. These valves are not intended for safety applications and dire consequences may result from using them in this role.

Check Valves

Check valves are important elements in any compression system. Single- and two-stage reciprocating compressors use a number of check valves in their systems. There is normally an input and output check valve in each pump head. These units will also incorporate an in-tank check valve, which is necessary during unload periods. A typical in-tank check valve (Fig. 7-9) is usually in the form of a National Pipe Thread (NPT) extension fitting. They have a ceramic poppet that is loaded against the seat with a spring. The spring and poppet are held in place with a spring deck and snap ring. Most in-tank check valves are equipped with a $1/8$-in. (F) NPT unloader port. These valves are available with either NPT input threads or compression fittings for use with copper tube.

Figure 7-10 shows an in-line check valve. The internals of these valves are similar to the in-tank valves and can generally be used in the same

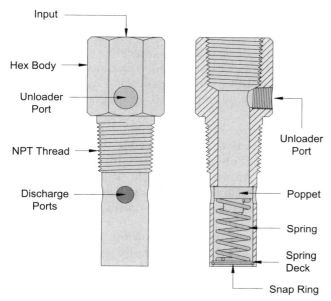

Figure 7-9 In-tank check valve. (*Courtesy of Control Devices, Inc.*)

applications. These valves also carry a $^1/_8$-in. (F) NPT port. Be cautious that the in-line check valve you select has a temperature rating that is suitable for compressor service. Do not select a flap type valve for compressed air service. These valves have the same outward appearance as the in-line check valve but are designed for liquid service and may not have the pressure and/or temperature rating required.

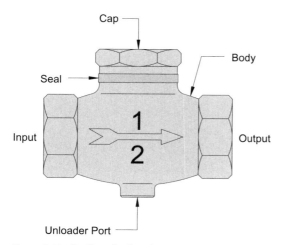

Figure 7-10 In-line check valve.

Expansion Joints and Vibration Isolators

Expansion/contraction and vibration are constant problems within the compression system. If left unaddressed, they can cause a multitude of problems throughout the system. Threaded joints can loosen and start to leak and, in extreme cases, these joints can crack. Copper and aluminum tubing are particularly susceptible to cracking, if installed improperly.

Two methods can be used to counter expansion/contraction and vibration. The first (Fig. 7-11) is to install flexible metal hose (flex joints) in areas that may be problematic. These joints are two short pieces of pipe connected with stainless steel bellows. The bellows are welded to the pipe stubs and a braided stainless steel jacket normally covers the bellows. Flex joints are readily available from many sources.

The second is to bend an expansion loop as shown in Fig. 7-12. Loops like this can be bent using pipe, but they are normally made from copper or aluminum tube. Great care should be taken to ensure that the loop has a sufficiently large radius, otherwise cracking could occur at the compression nuts.

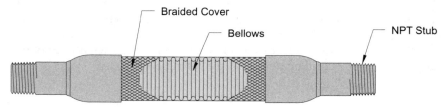

Braided Cover

Bellows

NPT Stub

Figure 7-11 Flexible metal hose.

Figure 7-12 Tube loop.

Hoses and Hose Connectors

In any compression system there is a requirement for hoses. Like everything else, many types of hoses are available, which are specifically designed for all sorts of industrial applications. When selecting compressed air hoses, great care should be taken to consider things like flexibility, abrasion resistance, UV protection, maximum operating temperature, pressure, exposure to oil and chemicals, SCFM, end connectors, and the like. Never use a hose for compressed air service that is not specifically rated for that duty. Never use a hose that is not clearly marked with its pressure and service rating, as shown in Fig. 7-13. If a hose has had all of its factory markings worn off, then it should be

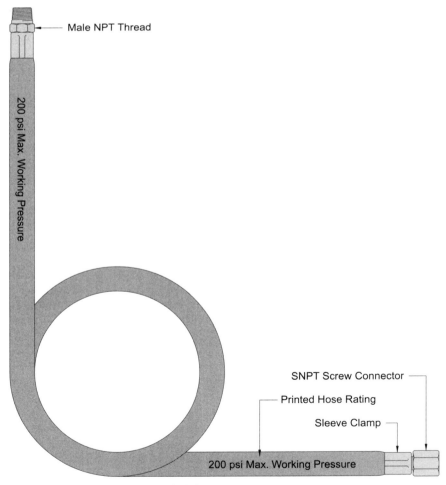

Figure 7-13 Standard compressed air hose.

removed from service and discarded. For general-purpose applications, standard hose assemblies can be purchased from a number of sources. These assemblies usually have a male NPT-threaded hose barb on one end and a straight NPT-threaded connector on the other end. Commercial hose assemblies are generally made up with sleeve clamps, as shown in Fig. 7-13.

Retractable Hoses

Overhead retractable hoses are rather convenient in many situations. Figure 7-14 shows a retractable hose reel that may be hung from the ceiling or the wall. These reels are very useful and can be seen in almost every auto repair shop in the nation. Figure 7-15 shows a coil hose. These hoses are perfect for assembly benches or applications that require only limited movement of the air tool. It should be noted that these hoses are a tangling nightmare if they are used while laying on a flat surface. Another attribute of retractable hoses is that they are extremely easy to put away.

Caution: When using air hoses, great care should be taken to ensure that the hose is never cut or disconnected while under pressure. Cutting or disconnecting an air hose under pressure will produce an extremely dangerous condition. The compressed air jetting from the end will cause the hose to "whip." A whipping air hose can break bones, damage equipment, smash

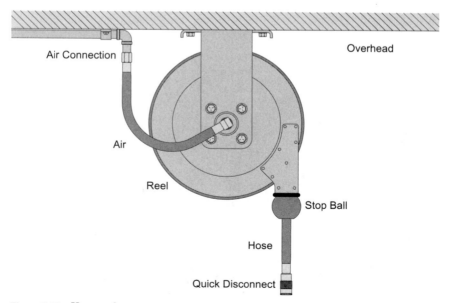

Figure 7-14 Hose reel.

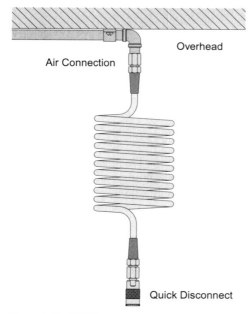

Figure 7-15   Coil hose.

doors, and shatter windows. I have witnessed this phenomenon on several occasions during the course of my career and can say with certainty that a whipping air hose has considerable destructive power.

In one instance in particular, I watched in horror as a 2-in. hose, with a universal hose connector on the end, bashed numerous football-sized dents into the side of a pickup truck. It seems that the hose had been disconnected for servicing but a lockout hadn't been placed on the compressor. When one of the shop personnel noticed that there wasn't any compressed air, he simply walked over and turned on the compressor … *a 350—horsepower compressor!*

Hose Clamps

Most commercial hose assemblies are delivered with sleeve-type hose clamps. These type of clamps are inexpensive, quick to assemble, and attractive. However, they do have a couple of drawbacks. They can't be tightened up when the assembly starts to leak, which leaves replacement as the only option. This brings to light their next problem—sleeve clamps can be very difficult to remove. In most instances they must be peeled off of the hose and fitting. The preferred replacement for the sleeve is the band clamp. These clamps are a band of metal that has two ears for crimping. The clamp is assembled with the hose and fitting using a special crimping tool to set the ears of the band. Band clamps provide excellent clamping force and are easily tightened and removed. The

Sleeve Clamp   Band Clamp   Screw Clamp

Figure 7-16   Hose clamps.

least desirable type of clamp is the screw clamp. These clamps are generally used for automotive maintenance. Screw clamps can be damaged easily by overtightening and the damage may not be apparent until the hose pops off the fitting while under pressure. Another drawback of screw clamps is that the end of the tab protrudes out from the assembly, which produces a cut hazard. If it is necessary to use a screw type hose clamp, then it should be wrapped with a protective cover of heavy industrial tape. Figure 7-16 shows the three types of hose clamps that are appropriate for compressed air hoses.

Hose Connectors

Along with hoses comes a requirement for hose connectors. There are a myriad of different types and sizes of hose connections that are perfectly suitable for compressed air. However, there are only few basic connectors that are typically used. Figures 7-17 through 7-20 show these hose connectors along with a brief description of each. Figure 7-17 shows two different male NPT × hose barb fittings. The combination nipple is a standard pipe nipple with a hose barb rolled on one end. These fittings are generally preferred for hoses larger than $1/2$ in. Hose menders are particularly useful items to have around and should be stocked in several different sizes.

Straight NPT screw connectors (Fig. 7-18) are a simple and effective way to connect basic compressed air hoses. These fittings are standard on one end of most $1/4$- and $3/8$-in. premade air hoses. The fitting consists of a ball seat/hose barb insert with a matching nut. The threads are straight national pipe threads and the seal is made against a small chamfer on the ID of the pipe. These fittings are available in sizes ranging from $1/4$ to $3/4$ in.

Industrial quick disconnect fittings are very common. They are extremely convenient to use because they automatically turn off the air when disconnected. This feature makes them ideal for applications that require frequent tool changes, such as auto repair shops. They are readily available in sizes ranging from $1/4$ to $3/4$ in. and can be purchased with male or female NPT threads and hose barbs. See Fig. 7-19.

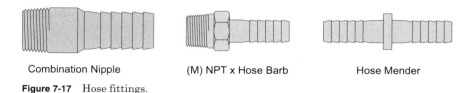

Combination Nipple   (M) NPT x Hose Barb   Hose Mender

Figure 7-17   Hose fittings.

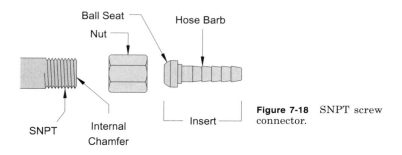

Figure 7-18   SNPT screw connector.

Figure 7-19   Industrial quick disconnect.

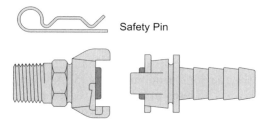

Figure 7-20   Universal hose coupling.

Universal hose couplings (Fig. 7-20) are typically used for larger compression systems. They are a quarter turn unisex design. Fittings in the $1/4$- to 1-in. range will universally mate with all sizes. This feature makes them rather popular in large plants and field operations. Additionally, these fittings are very rugged and easy to service. They are available in sizes up to 2 in., but these larger sizes will not mate with the smaller fittings. The one significant drawback to these fittings is that they do

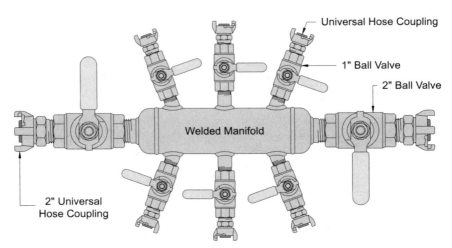

Figure 7-21   Universal hose connector manifold.

not turn off the air when disconnected. These fittings must be installed in conjunction with a ball valve. To reduce the possibility of accidental disconnection while under pressure, these fittings should always be secured with a safety pin when connected. Figure 7-21 shows a manifold designed to support large field operations. The primary distribution loop is laid out with 2-in. hose sections that are interrupted with these manifolds. The manifolds provide convenient points to which workmen can connect their applications.

Drain Valves

Water buildup is a constant problem within compressed air systems. Draining accumulated water is a must, and more often than not it is neglected. This can lead to some rather serious problems. If neglected long enough, water can seriously reduce the capacity of your receiver and completely fill your piping. This has the devastating effect of delivering water instead of air to your pneumatic tools. Most pieces of compression equipment are equipped with drain valves; however, the valves provided are usually the least expensive unit that is available at the time of manufacture. It makes a lot of sense to replace these low-cost units with better quality valves.

Automatic drains are an excellent addition to any compressed air system. Figure 7-22 shows an automatic electronic drain. These types of drains are the preferred solution for most compressor systems. They typically operate on line voltage or may be wired into the compressor motor controls. They have two controls: an "on time" and an "interval time." The on time is the amount of time that the valve is open and the

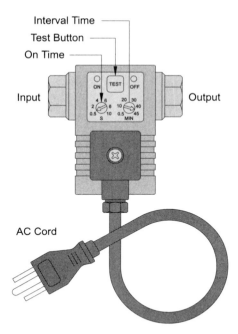

Interval Time
Test Button
On Time
Input
Output
AC Cord

Figure 7-22 Electronic drain.

interval time is how often the valve opens. On times are normally in seconds and interval times are normally in minutes.

Figure 7-23 is a schematic representation of a float drain. These drains are usually used in applications that do not have access to AC power. Because of their delicate nature, float drains are not particularly reliable and require regular service.

Most compressors are delivered with an external seat drain valve as shown in Fig. 7-24. These are the same valves that are used on most

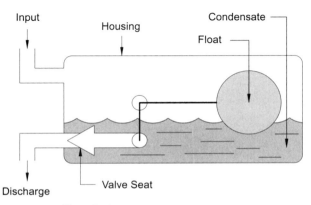

Input
Housing
Condensate
Float
Discharge
Valve Seat

Figure 7-23 Float drain.

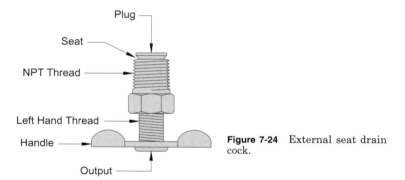

Figure 7-24 External seat drain cock.

automotive radiators. They have a body that has an internal left hand thread and an external NPT thread. The threaded end of the body has a seat on the ID. The plug has a matching external left hand thread and a tapered head that engages the seat. When the plug is screwed in, a port is exposed and the water drains through the output. Generally speaking, these are particularly inferior devices and should be removed and discarded as soon as you receive your compressor. They should be replaced with a quality manual valve or an automatic electronic drain with strainer.

A plug valve (Fig. 7-25) is a much better selection for a drain. These valves are inexpensive, reliable, and easy to operate. They typically have a brass body with a tapered plug as the valve element. The plug is held in constant engagement with a preload spring. They are available with either male or female NPT threads.

Cable drain valves (Fig. 7-26) are commonly found on truck air brake systems. However, they can be used as a handy drain valve in other applications. The drain process is accomplished by simply tugging on the attached cable, which can be routed to a convenient location. These valves are particularly handy solutions for applications that only require occasional draining.

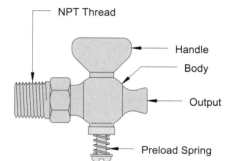

Figure 7-25 Plug valve.

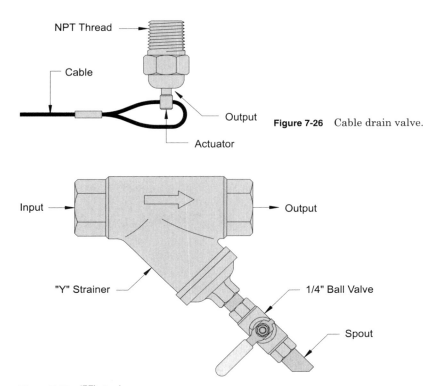

Figure 7-26 Cable drain valve.

Figure 7-27 "Y" strainer.

Regardless of the type of drain that is used, a strainer element should be installed to protect the valve from the debris, which moves through any compression system. A "Y" type strainer may be used, as shown in Fig. 7-27. The clean out port should be equipped with a ball valve so that collected debris can be periodically blown out. If a "Y" strainer is used, be certain to select an element that is sufficiently fine to protect the drain valve.

The preferred strainer for compressed air applications is shown in Fig. 7-28. These strainers are specifically designed for compression systems. The strainer carries a ball valve, which is used to isolate the strainer element. When the valve is turned off, the element may be removed and cleaned without disturbing the operation of the compression system.

Commercial Pipe Fittings

There are a myriad of different types of pipe and tube fittings that are perfectly acceptable for use with compressed air. However, it's a good idea to stick with readily available NPT and compression fittings when working with a distribution system. This will make future service and

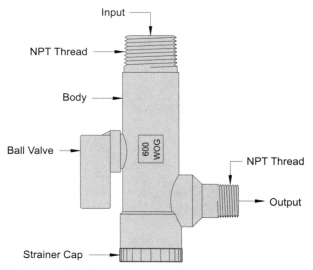

Figure 7-28 Strainer.

expansion considerably easier. Using less common or special fittings can create the problem of shutting down an entire compression system while a replacement part is overnighted from out of state.

It is also a good idea to have a clear understanding of the different classes of fittings. For residential water and gas applications, 125 psi class is generally used. These are the galvanized fittings that you find at the hardware store or home improvement center. This class of fittings should not be used for commercial compressed air systems. For compression systems with a working pressure of 150 psi or less, 150 psi class fittings, or what are sometimes called malleable iron, are the preferred choice. If the system pressure is higher than 150 psi, then 300 psi class fittings must be used. These fittings are also malleable iron but are considerably heavier than their 150 psi counterparts. Most petrochemical plants, refineries, and marine applications specify 3000 psi class fittings, or what are sometimes referred to as forged steel. The cost of 300 and 3000 psi class fittings can be prohibitive and they should not be used unless the system pressure or customer specifications call for them.

Figure 7-29 shows some commercial NPT fittings that are readily available. All of these fittings are available in 150, 300, and 3000 psi classes. A close nipple (shown) is the shortest piece of pipe that can have a full taper thread on both ends. Do not use "fully threaded" nipples. These are usually a piece of pipe with a straight thread and they will not seal properly. Similarly, do not use merchant couplings. Merchant couplings have a straight internal thread and they will not seal properly. Fully threaded nipples and merchant couplings are generally used for electrical conduit or low-pressure applications.

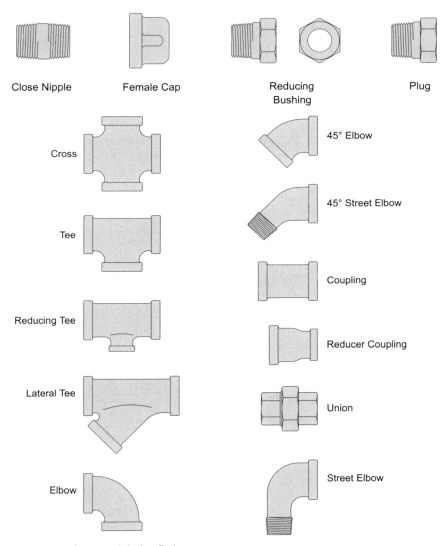

Figure 7-29   Commercial pipe fittings.

Compression Fittings

For lesser plumbing applications, standard compression fittings are ideal. These fittings are very useful for instrumentation, control, and small compressors. The tube to be connected is inserted through the nut and a ferrule is slid onto the tube. The nut is then screwed onto the body. As the nut is tightened, the ferrule is deformed into the OD of the tube and forms a swaged seal. Figure 7-30 shows a compression fitting assembly. These

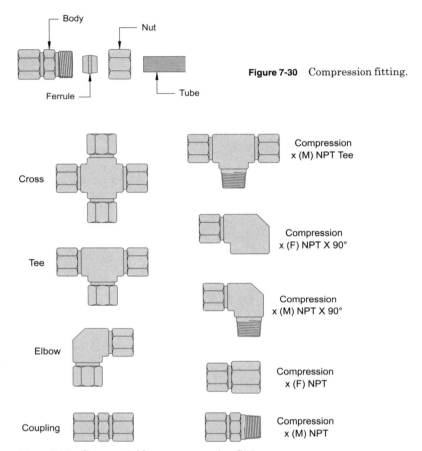

Figure 7-30   Compression fitting.

Figure 7-31   Commercial brass compression fittings.

fittings are generally available in all the same patterns as pipe fittings. Figure 7-31 show the more common commercial compression fittings.

Pipe Hangers

It is very important to properly support installed piping. Improperly supported pipe can result in the transmission of excessive noise into the frame of the building, reoccurring leaks, cracking, and, in extreme cases, buckling. Pipe should be supported by some type of commercial pipe hangers. There are a variety of systems on the market, and any one can be effective in this role. In most cases, some combination of hanger types is used.

Figure 7-32 shows a typical pipe hanger. The pipe rests in a band-type saddle that is supported with a bridge and hanger bolt. The bridge is

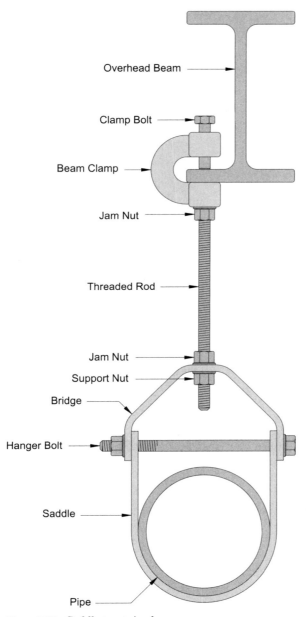

Overhead Beam

Clamp Bolt

Beam Clamp

Jam Nut

Threaded Rod

Jam Nut

Support Nut

Bridge

Hanger Bolt

Saddle

Pipe

Figure 7-32   Saddle type pipe hanger.

supported via a threaded rod that screws into a beam clamp. The beam clamp is tightened onto the flange of an overhead beam.

This type of hanger is particularly effective because of the degree of freedom that it provides. The long rod allows the pipe to "float," which will compensate for expansion and contraction. It will also help eliminate the transfer of noise into the frame of the building. Additionally, the length of the rod can be adjusted to control the incline, or level, of the pipe.

Ball Valves

Ball valves are the most user-friendly control elements in a compression system. These valves are inexpensive, reliable, and easy to operate. They are available in a myriad of patterns, pressure ratings, materials, and mounting methods. The basic ball valve consists of a body with a ball imbedded in the center. The ball is drilled through on center and when it is rotated 90°, the hole aligns with the bore of the body, opening the valve. When the ball is rotated back, the valve is closed. Figure 7-33 shows a schematic representation of a ball valve. Figure 7-34 shows a typical commercial ball valve. This particular unit has two female NPT threads; however, ball valves are also available in male/female or male/male configurations.

Three-way valves allow the input to be connected to either output A or output B. The center position generally closes the valve. Figure 7-35 shows a typical three-way ball valve that might be used to control a spring return cylinder.

Four-way valves are similar in operation to a three-way valve, except that these units have a venting function. When output A is open to the input, output B is open to the vent port. When output B is open to the input, output A is open to the vent port. Figure 7-36 shows a typical four-way venting valve configuration. This type of valve may be used to control a standard cylinder or as a direction control for an air motor.

Gate Valves

In larger sizes, ball valves may become prohibitively expensive; therefore, gate valves are usually the choice in these situations. The gate valve (Fig. 7-37) consists of a body with a sliding gate. To open the valve the gate is lifted out of the bore by turning a threaded actuator. To close the valve the gate is replaced in the bore by an opposite rotation. Gate valves are designed to provide an on/off function and should not be used for

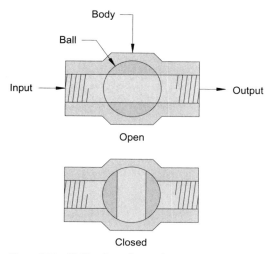

Figure 7-33 Ball valve schematic.

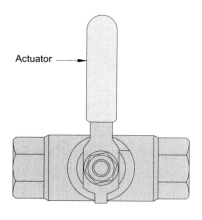

Figure 7-34 Commercial ball valve.

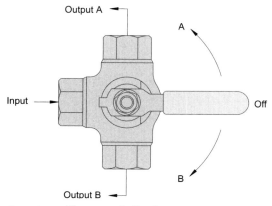

Figure 7-35 Three-way ball valve.

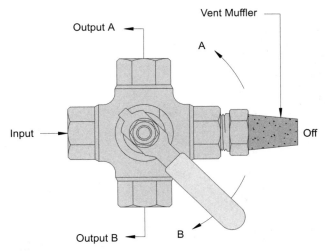

Figure 7-36   Four-way ball valve.

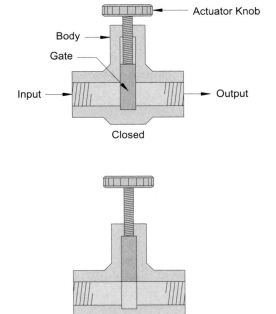

Figure 7-37   Gate valve schematic.

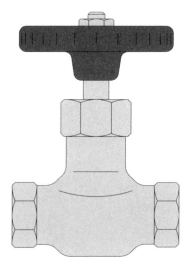

Figure 7-38   Gate valve.

applications that require flow control. These valves are normally used when sizes become too large to economically use ball valves. Figure 7-38 shows a typical commercial gate valve. The valve is shown with two female NPT threads; however, these valves are commonly available with standard flanges.

Globe Valves

The least expensive design of the three valve types is the globe valve. This valve is similar in application, operation, and appearance to the gate valve. The principal difference is the sealing mechanism. By rotating the actuator knob, the seat is forced into the bore and the valve closes. To open the valve the actuator is counter-rotated and the seat is lifted. Figure 7-39 shows schematic representation of a globe valve. Figure 7-40 shows a typical commercial globe valve. These valves are commonly found in residential water service and generally do not represent good choice for compressed air service.

Butterfly Valves

Butterfly valves are generally low-cost alternatives to other designs. These valves offer low flow restriction in a compact package. They are not considered to be particularly durable designs and are generally used in situations that do not require frequent actuation. Most butterfly

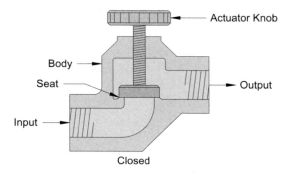

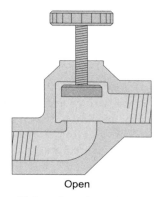

Figure 7-39 Globe valve schematic.

Figure 7-40 Globe valve.

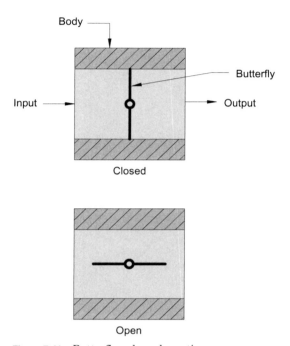

Figure 7-41   Butterfly valve schematic.

valves are built into a body that matches standard flange patterns. Because of their compact design, they can be placed in between two existing flanges by simply spreading the joint and inserting the valve. This particular attribute makes them rather popular with maintenance departments. Figure 7-41 shows a schematic representation of a butterfly valve. Figure 7-42 shows a typical commercial butterfly valve. Note the lock handle on the actuator. Butterfly valves have a tendency to "flutter" in high flow situations. To combat this tendency, these valves are normally supplied with some sort of locking system.

Solenoid Valves

Solenoid valves are electrically actuated valves. These valves are very common in pneumatic control systems. They are available in a variety of voltages, sizes, and flow rates. Figure 7-43 shows a typical two-way solenoid valve. These units are usually a body with a replaceable coil. The coils may be had with junction boxes, plug type connectors, straight leads, or conduit terminations. The bodies are commonly available in two-, three-, or four-way patterns.

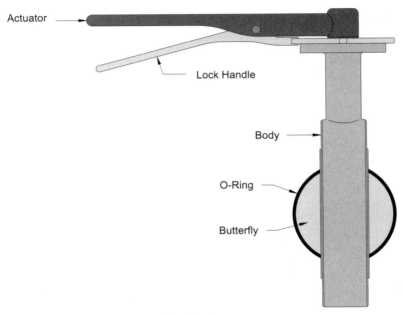

Figure 7-42   Commercial butterfly valve.

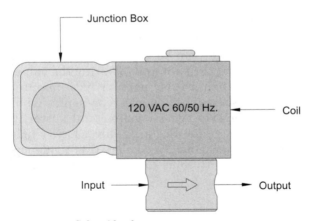

Figure 7-43   Solenoid valve.

Pilot Valves

Pilot valves serve the same basic function as a solenoid valve, except they are actuated by a pressure feed rather than an electrical signal. Not as common as their electrical counterparts, these valves often find favor in complex pneumatic circuitry. They are available in a variety of sizes

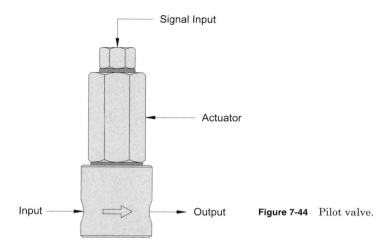

Signal Input

Actuator

Input   Output   **Figure 7-44**   Pilot valve.

and flow rates, but they are normally found in sizes applicable for pneumatic circuitry. Figure 7-44 shows a typical pilot valve.

Needle Valves

Needle valves are used to restrict and control flow rates. These valves normally consist of a body with a tapered needle within a seat. To adjust the flow rate, the needle is screwed in or out via the adjustment knob. Needle valves are very common and are available in different sizes, flow rates, sensitivities, and ports. Figure 7-45 shows a schematic representation of a needle valve. The needle is adjusted in and out of the seat to change the effective orifice size. Figure 7-46 shows a typical 90° needle valve with female NPT ports.

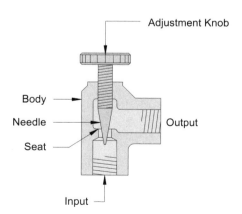

Adjustment Knob

Body

Needle

Seat

Output

Figure 7-45   Needle valve schematic.

Input

Adjustment Knob

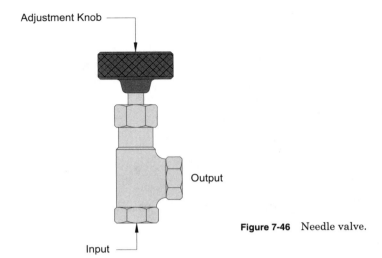

Output

Figure 7-46   Needle valve.

Input

Speed Control Valves

Speed control valves (also discussed in Chap. 2) are primarily used to control the rate at which cylinders extend and retract. These valves typically consist of a check valve and needle valve placed in a parallel configuration. The check valve allows full flow in one direction while forcing the flow through the needle valve in the other direction. The speed of the cylinder (flow rate) is adjusted via the needle valve. Figure 7-47 shows a typical speed control valve.

Speed Adjustment

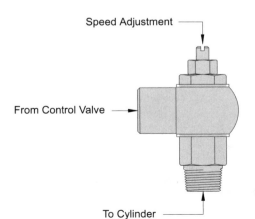

From Control Valve

Figure 7-47   Speed control valve.

To Cylinder

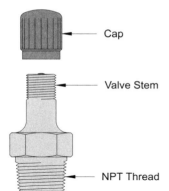

Cap

Valve Stem

Figure 7-48   Tire valve fitting.

NPT Thread

Tire Valve Fittings

Tire valve fittings, or Schrader valves as they are referred to, can be particularly useful for applications that require charging with compressed air, but that do not need to be connected to the system. A good example of this is water charged fire extinguishers. These units are filled with water and then sealed up. After sealing the body of the extinguisher, it is charged with compressed air by simply using a tire filler. Other uses include charging pneumatic springs, dampers, portable compressed air tanks, and hydraulic accumulators. Figure 7-48 shows a typical Schrader valve with a male NPT thread.

Mufflers

Escaping compressed air can be very loud, in many instances loud enough to produce a safety hazard. Applying a muffler will help defuse the exiting air and mitigate the noise level. Mufflers come in many sizes and attenuation levels. One of the most common mufflers is the sintered element unit. These devices are an NPT fitting with a diffuser element made by pressing small bronze balls together. After the pressing process, small spaces remain between the balls, which allow air to escape. Another common muffler is a screen diffuser. The air is allowed to enter a chamber, which is surrounded with a fine mesh screen. The screen serves to break up the flow and diffuse the air. Figure 7-49 shows both sintered element and screen-type mufflers.

Intake Air Filters

For obvious reasons it is a good idea to filter the incoming air to the compressor pump. The environment in which compressors are forced to operate are typically very dirty, making a high-quality air filter necessary

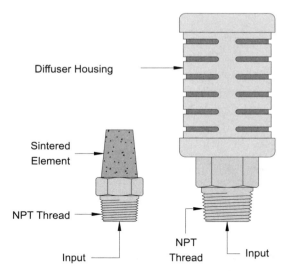

Diffuser Housing

Sintered
Element

NPT Thread

Input

NPT
Thread

Input

Figure 7-49   Pneumatic muffler.

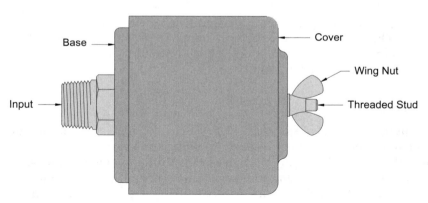

Base

Cover

Wing Nut

Input

Threaded Stud

Figure 7-50   Intake air filter.

to protect the equipment. Figure 7-50 shows a typical intake air filter. These units use a paper element similar in construction to an ordinary automotive air filter. The elements are very inexpensive and should be replaced on a regular basis.

Pressure Gauges

It is absolutely mandatory to monitor pressures within a compressed air system. Pressure indicators are critical for gauging the performance of

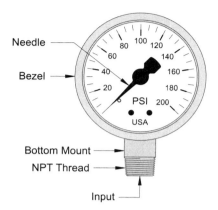

Needle

Bezel

Bottom Mount

NPT Thread

Input

Figure 7-51 Bottom mount pressure gauge.

your compressor and your applications. To accomplish this, pressure gauges are used. These units are available in a wide variety of sizes, ratings, materials, and configurations. Figure 7-51 shows a typical pressure gauge that may be found on any compressor. This particular unit is a bottom mount while Fig. 7-52 shows a rear mount unit.

Figure 7-53 shows a schematic representation of the internals of a Borden tube pressure gauge. The pressure to be monitored is introduced into the input. The Borden tube is soldered into the input fitting. The Borden tube is a flattened tube, which has been bent into a circular arc. The far end is sealed and equipped with a pivot point. The pivot is connected to a circular gear rack via a toggle link. The circular rack drives a pinion gear, which is connected to the indicator needle. As the pressure builds within the tube, it tries to straighten out and the pivot end moves. As the pivot end moves, it pulls on the toggle link, which rotates the circular rack. As the rack rotates, it forces rotation of the pinion gear and the indicator needle shows a corresponding pressure reading.

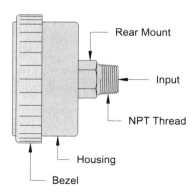

Rear Mount

Input

NPT Thread

Housing

Bezel

Figure 7-52 Rear mount pressure gauge.

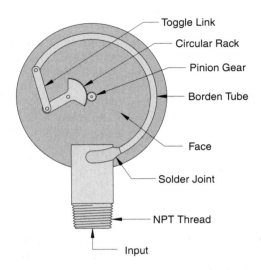

Toggle Link

Circular Rack

Pinion Gear

Borden Tube

Face

Solder Joint

NPT Thread

Input

Figure 7-53 Gauge internals.

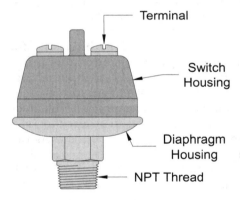

Terminal

Switch
Housing

Diaphragm
Housing

NPT Thread

Figure 7-54 Pressure switch.

Pressure Switches

Pressure switches are devices that open or close an electrical switch at a predetermined setting. These switches can be delivered with either fixed pressure settings or adjustable settings. They are commonly available with pressure ratings from vacuum to several thousand psi. The switches are normally rated for fairly low currents, which prohibits them from switching high electrical loads directly. Figure 7-54 shows a typical fixed setting pressure switch.

Compressor Control Switches

Taking the pressure switch a step further is the compressor control switch (Fig. 7-55). These are normally adjustable differential pressure

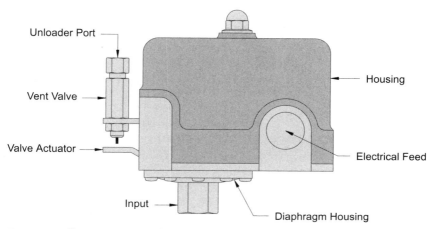

Figure 7-55   Compressor control switch.

switches with an unloader function. The switches will usually be a double pole, single throw (DPST) unit with a high-pressure (cut off) setting and a low-pressure (cut on) setting. The settings are adjusted to control the pressure at which the compressor turns on and off. When the pressure switch hits its high limit (cut off) the switch opens, disconnecting power to the motor controller. At the same time that the switch opens it also actuates a vent valve, which is normally mounted on the side of the switch housing. The vent valve is connected to the output of the pump and bleeds down the head pressure, or unloads, when the compressor shuts off.

Continuous-Run Unloader Valves

For compressors that must run continuously, such as engine-driven or high duty cycle units, it is necessary to unload the pump while it is rotating. To accomplish this, a continuous-run unloader valve is used. At a preset upper limit, the valve will connect its input port to the vent port and allow the pump to vent to atmosphere. When the valve sees a preset lower limit the valve closes the vent and the pump output is directed through the output port and into the receiver. These valves typically have a manual unloader control that is utilized for start-up operations. For engine-driven compressors, a throttle control can be added to these units. When the valve unloads, pressure is applied to the throttle cylinder and the engine is pulled down to idle speed. When the valve loads, the cylinder pressure is vented and an integral spring opens the throttle to run speed. Figure 7-56 shows a common continuous-run unloader valve with throttle control cylinder.

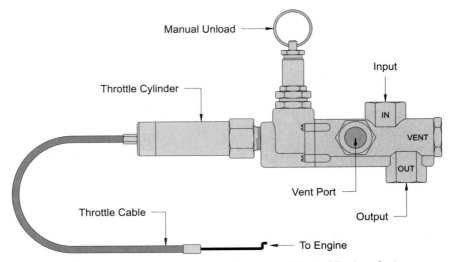

Figure 7-56   Continuous-run unloader valve. (*Courtesy of Control Devices, Inc.*)

Vee Belts

In general, a compressor's motor and pump are connected through a standard vee belt. Smaller compressors use a single belt, while larger units normally use belt sets. When replacing the latter, it is important to purchase matched sets. Never replace only one belt in a multi-belt set. The reason for this is that the belt lengths differ slightly and when unmatched belts are used, one of them will carry the bulk of the load. This, in turn, will accelerate the wear on that particular belt.

Figure 7-57 and the formula below can be used to determine the approximate length for replacement belts.

$$L = (2 \times P) \div 1.57 \times (C \div M)$$

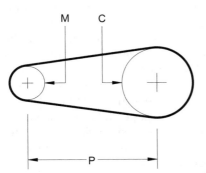

Figure 7-57   Determining belt length.

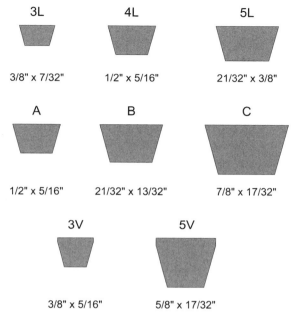

3L	4L	5L
3/8" x 7/32"	1/2" x 5/16"	21/32" x 3/8"
A	B	C
1/2" x 5/16"	21/32" x 13/32"	7/8" x 17/32"
3V	5V
3/8" x 5/16"	5/8" x 17/32"

Figure 7-58   Common commercial v-belt sizes.

where L = pitch length of belt
P = center distance
C = pump sheave diameter
M = motor sheave diameter

Figure 7-58 shows the various sizes of commercial vee belts. $3L$, $4L$, and $5L$ belts are normally reserved for general purpose applications, although they are used extensively on small compressors. A, B, and C belts are the most common belt found in industrial applications. These belts are readily available in matched sets and, when maintained properly, will provide many years of service. $3V$ and $5V$ belts are normally found in heavy-duty power transmission service.

Sound Dampening Enclosures

Most compressor manufacturers offer optional sound dampening enclosures for their standard compressor packages. When specifying a compressor, only include a sound dampening enclosure if absolutely necessary. Although these housings may seem like a good idea when the compressor is new, they become a considerable hindrance during maintenance periods. The housings severely limit access to the compression equipment,

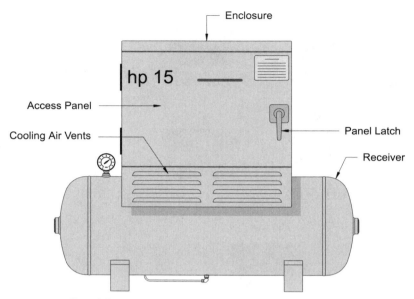

Figure 7-59 Sound damping enclosure.

which translates to considerable difficulty during scheduled maintenance. Additionally, the enclosures greatly limit the ability to conduct
routine maintenance. After a few years of operation, the panels of most
enclosures are nowhere to be found. However, the original framework
is still in place and doing a fine job of hindering maintenance operations.
Figure 7-59 shows a typical sound dampening enclosure on a small
packaged compressor.

Compressor Mounts

Larger compressors are typically supplied with a suitable mounting
system that is integral to the frame. However, smaller packaged compressors are normally delivered with rigid mounts. A compressor should
never be mounted directly to a concrete floor without some sort of
resilient mount. This can cause cracking around the welds that hold the
feet and motor/pump mount to the receiver. There is a dizzying array
of vibration isolators available on the market. The problem is that it is
almost impossible to select a suitable mount without considerable
research. Building a simple and effective compressor mount is not a
difficult proposition. Figure 7-60 shows two illustrations that outline the
construction of these mounts. Mounts of this nature are perfectly suited
for compressors in the 5- to 50-hp range and can be constructed for only

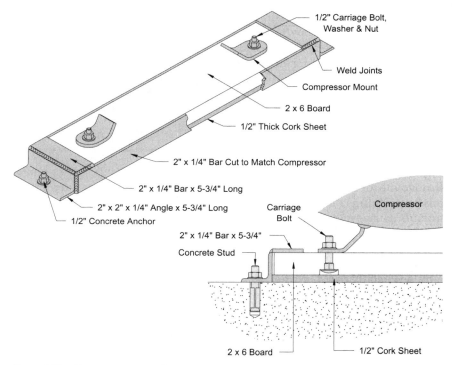

Figure 7-60   Compressor mount.

a few dollars. The mount is simply a 2×6 that has a $^1/_2$-in. cork pad glued to the bottom. The 2×6 is bolted to the compressor feet and in turn held in place with a steel frame that is bolted to the floor.

Skid Mounting

For some field operations the compression system is mounted on a custom-built equipment skid. To simplify matters at the application sight, the end user will generally specify a compression package that can be lifted into position, interconnected and turned on. Skid mounted systems are often used on ships, offshore drilling and production platforms, mining operations, chemical plants, drilling sites, and the like.

Figure 7-61 shows a typical redundant compression system that has been configured on an equipment skid. The skid will typically incorporate four lifting eyes and a lip around the entire deck area. The lip will catch any fluid contaminates that might leak from the system.

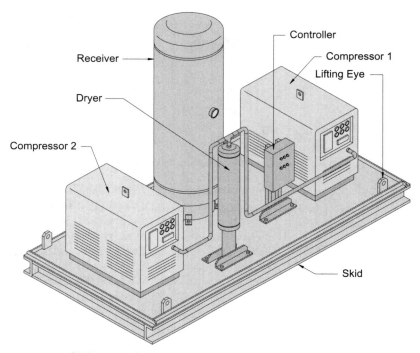

Figure 7-61 Skid mounted compression system.

The skids are normally fabricated from standard steel shapes. Fluid lips are usually constructed using 2 in. × 2 in. × $\frac{1}{4}$ in. angle and the deck plate is $\frac{1}{4}$-in.-thick tread plate. The actual frame is constructed from either channel or "I" beam that has been specifically selected to support the overall weight of the assembled system during lifting operations. All major equipment should be bolted to one of the beams, while minor equipment may be through-bolted to the deck plate. Figure 7-62 shows an exploded view of a typical equipment skid.

The heavy equipment may be mounted by through-drilling the flanges of the rails. However, many engineers will not allow through holes in the flanges of the structural components of a skid. They will opt instead for a mounting pad, as shown in Fig. 7-63. The pad is usually a $\frac{3}{4}$-in. plate with a tapped hole on center. The pad is then welded into position on a corresponding rail. A threaded stud is placed into the hole and the equipment is secured with a heavy nut and washer.

The fluid lip is generally fabricated with 2 in. × 2 in. × $\frac{1}{4}$ in. steel angle, as shown in Fig. 7-64. Note the $\frac{1}{4}$-in. offset on the edge of the skid. This is intended to allow a lap weld joint to ease the welding effort and

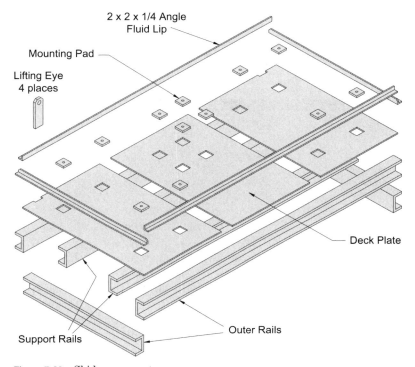

Figure 7-62 Skid components.

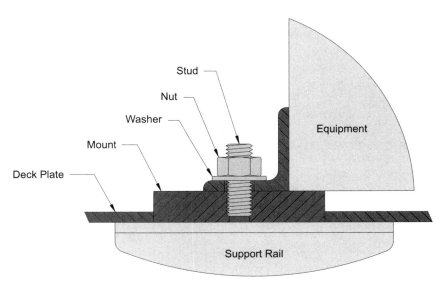

Figure 7-63 Mount pad.

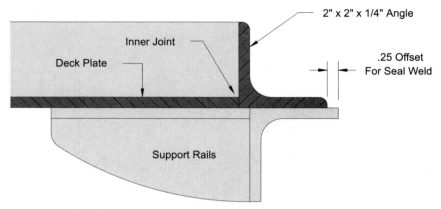

Figure 7-64   Fluid lip detail.

improve appearance. Typically, a seal weld is specified for both the outer and inner joints of the lip.

Facilities must be provided for draining the skid in the event of a spill. Generally one or two drain ports are incorporated into the outer perimeter of the skid. Figure 7-65 shows two of the most common methods for installing these drains. The best method is to plumb an elbow and two short pipe stubs to a half-coupling, which is welded into the web of the outer rail. When welded properly, this arrangement will not adversely affect the strength of the beam. However, some engineers are reluctant to make any penetrations in the rails. In these cases, a half-coupling is welded into the fluid lip, as shown. It should be noted that a drain should not be routed through the flange of any of the rails.

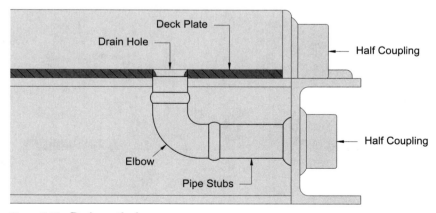

Figure 7-65   Drain methods.

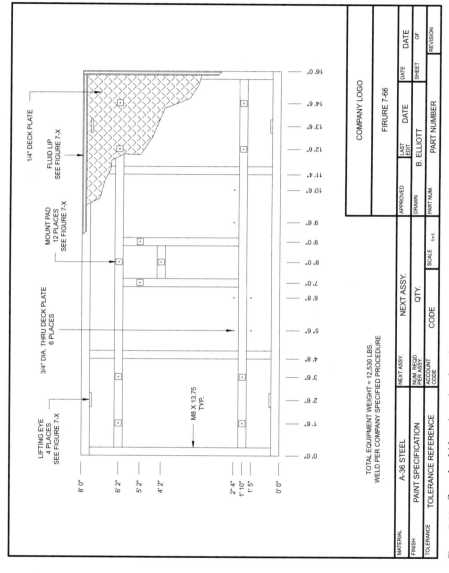

Figure 7-66 Sample skid engineering drawing.

When installing the piping on an equipment skid, it is important to consider expansion and contraction. The close proximity of the equipment creates two situations that can produce damaging loads on the equipment if the plumbing is not designed and installed properly. The first situation is encountered during lifting operations. Whenever an equipment skid is lifted, there will invariably be some flexing of the frame. If this flexing is not taken into account, the plumbing can place very high loads on the mounted equipment. These loads can buckle receivers, crack aftercoolers, and create leaks in threaded and flanged joints. The second situation is thermal expansion and contraction. Thermal considerations are not as critical as lifting requirements because the piping is generally made from the same material as the skid, so both items expand and contract at the same rate. The real thermal consideration is with the output temperature of the compressor, which is usually much higher than ambient. Figure 7-67 shows an incorrect connection, while Fig. 7-68 shows how the connection should be routed to allow for any flexure.

Figure 7-67   Incorrect expansion connection.

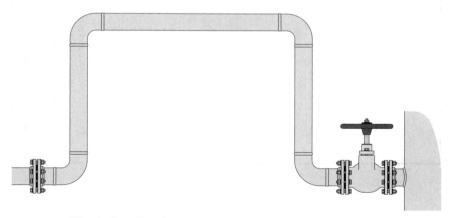

Figure 7-68   Plumbed to allow for expansion.

Questions*

1. Which is the most important regulator in the compression system?
   (A) Paint gun   (B) Master   (C) Receiver   (D) Line

2. What type of service should an operator perform on a safety valve?
   (A) Replacement   (B) None   (C) Rebuild   (D) Pressure setting

3. What is the poppet made of in an in-tank check valve?
   (A) Plastic   (B) Stainless steel   (C) Brass   (D) Ceramic

4. What is the best way to deal with expansion and vibration in piping?
   (A) Hose   (B) Loops   (C) Loose fittings   (D) Flex joints

5. Name two common hose connectors?
   (A) Cam lock   (B) Garden hose   (C) Universal
   (D) Quick disconnect

6. What is the principal danger when dealing with hoses?
   (A) Leaks   (B) Whip   (C) Weight   (D) Dirt

7. What is an important accessory for automatic drains?
   (A) Paint   (B) Teflon tape   (C) Strainer   (D) O-ring

8. What additional function does the pressure switch usually have?
   (A) Unloading   (B) Alarm   (C) Timer   (D) Motor controller

9. Name two common valves.
   (A) Ball   (B) Slide   (C) Gate   (D) Motor

10. What is the most common pressure gauge type?
    (A) Diaphragm   (B) Borden tube   (C) Piston   (D) Liquid

*Circle all that apply.

Pneumatic Controls

The intention of this book is to review air compression systems and not necessarily control systems. However, during the course of your compressed air experience, you will likely be called upon to deal with some pneumatic control systems. Even basic air tools have simple control systems that can render the equipment useless if they fail. Auto lifts, for example, have a very basic control system that is not intuitively understandable in any way. If a malfunction occurs, the shop bay is useless until the lift is repaired. Any average auto mechanic is capable of repairing the controls on the lift in just a few minutes, if he has some basic knowledge of pneumatics. Pressure amplifiers are another example. They have simple, basic controls that are not intuitive in appearance. Rather than shutting down an expensive machine tool to wait for a repair technician, a machinist with a basic understanding of pneumatics could easily fix the problem. Having a cursory exposure to pneumatic symbols and equivalent assemblies will go a long way to get you started in this arena.

Table 8-1 shows some of the more common pneumatic symbols that are used in the industry. After reviewing these symbols, it will become obvious that they are not as clear as one might want. To make matters worse, companies use modified versions of these symbols or they use symbols that have been designed in-house. Most of these "nonstandard" symbols are actually intended to make technical drawings easier to interpret for the layman.

TABLE 8-1   Pneumatic Symbols

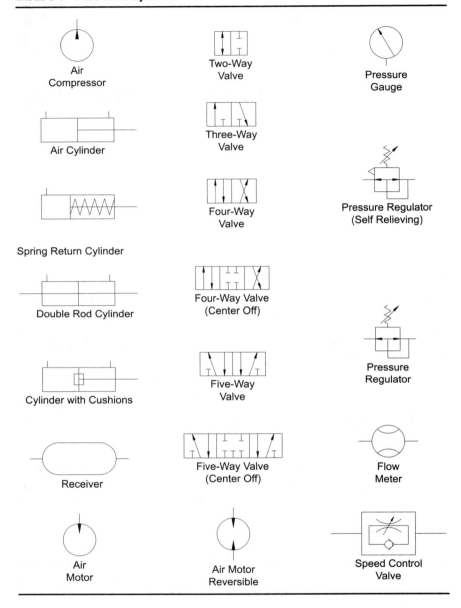

Air
Compressor

Air Cylinder

Spring Return Cylinder

Double Rod Cylinder

Cylinder with Cushions

Receiver

Air
Motor

Two-Way
Valve

Three-Way
Valve

Four-Way
Valve

Four-Way Valve
(Center Off)

Five-Way
Valve

Five-Way Valve
(Center Off)

Air Motor
Reversible

Pressure
Gauge

Pressure Regulator
(Self Relieving)

Pressure
Regulator

Flow
Meter

Speed Control
Valve

TABLE 8-1 Pneumatic Symbols (*Continued*)

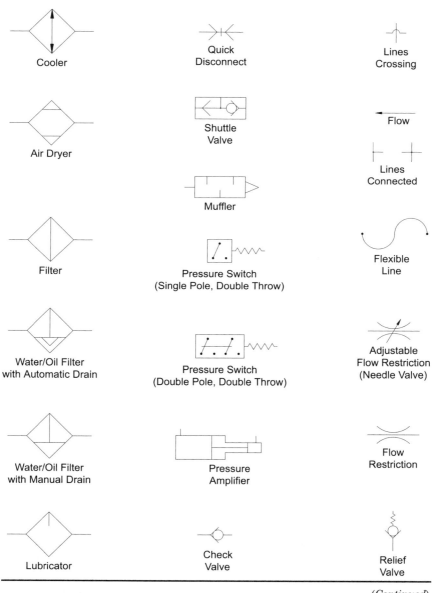

Cooler	Quick Disconnect	Lines Crossing
Air Dryer	Shuttle Valve	Flow
	Muffler	Lines Connected
Filter	Pressure Switch (Single Pole, Double Throw)	Flexible Line
Water/Oil Filter with Automatic Drain	Pressure Switch (Double Pole, Double Throw)	Adjustable Flow Restriction (Needle Valve)
Water/Oil Filter with Manual Drain	Pressure Amplifier	Flow Restriction
Lubricator	Check Valve	Relief Valve

(*Continued*)

TABLE 8-1   Pneumatic Symbols (*Continued*)

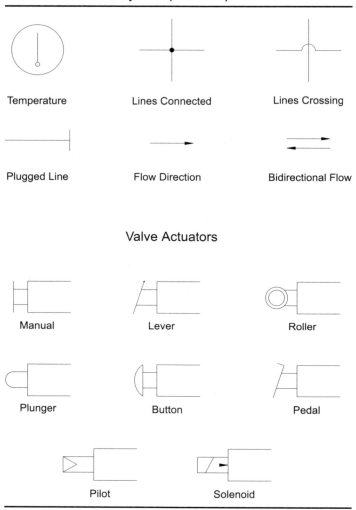

Temperature	Lines Connected	Lines Crossing
Plugged Line	Flow Direction	Bidirectional Flow

Valve Actuators

Manual	Lever	Roller
Plunger	Button	Pedal
	Pilot	Solenoid

One of the most important controls in a pneumatic system is the regulator. Figure 8-1 shows a cutaway view of a nonventing diaphragm regulator. For more information on the operation of these devices refer to Chap. 7.

Figure 8-2 shows a typical workstation as may be found in an automobile paint booth. System air is fed through a valve, coalescing filter, desiccant dryer, particulate filter, and regulator. The regulator is used to adjust the feed pressure to an appropriate level. The output of the regulator is equipped with an industrial quick disconnect and hose. Figure 8-3 shows how a schematic representation of the workstation might appear on a technical drawing.

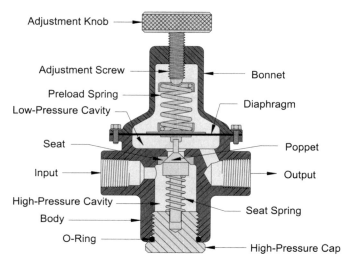

Figure 8-1 Pressure regulator.

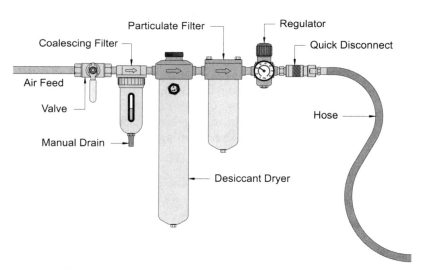

Figure 8-2 Workstation.

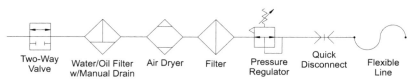

Figure 8-3 Schematic representation.

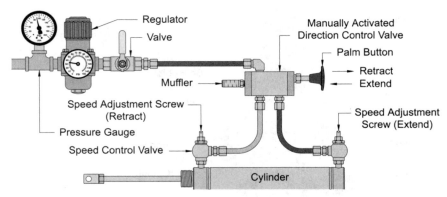

Figure 8-4 Cylinder control system.

Figure 8-4 shows a typical pneumatic cylinder control system. System air is fed through a regulator and valve. The regulator is used to tune the cylinder force. The output of the regulator is directed into a four-way venting control valve, which is used to extend or retract the cylinder. The output of the valve is connected to the cylinder through two speed control valves. These valves control the rates at which the cylinder will extend and contract. Figure 8-5 shows how a schematic representation of this assembly might appear on a technical drawing.

Figure 8-6 shows a typical screw compression system. The system has a screw compressor, which feeds an after cooler. There is a water trap on the output of the aftercooler with a manual drain. The output of the

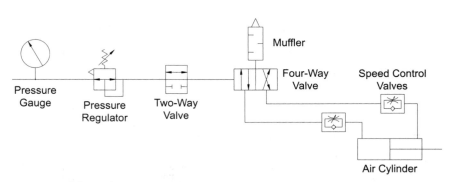

Figure 8-5 Schematic representation.

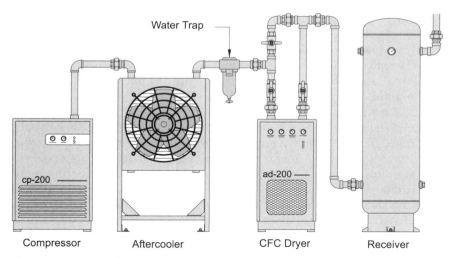

Figure 8-6 Compression system.

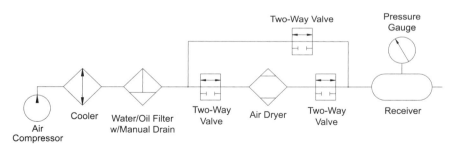

Figure 8-7 Schematic representation.

trap is connected to a bypass manifold set, intended to isolate the dryer for service. The bypass manifold also serves as an interface between the water trap, dryer, and receiver. Figure 8-7 shows how this compression system might appear on a technical drawing.

In-House Pneumatic Symbols

For many companies the use of pneumatic controls may be limited to just a few applications and, therefore, a full-time pneumatics engineer is not warranted. The duties of designing and modifying pneumatic equipment generally falls on the mechanical engineer, or whoever else might volunteer for the task. Similarly, maintenance personnel may be highly trained and skilled in certain equipment and disciplines but may have very little knowledge of pneumatics. Nonetheless, these personnel will be called on to repair and modify the limited pneumatic equipment in the plant.

Standard pneumatic symbols are confusing enough for trained personnel. When untrained personnel are forced to deal with these symbols, the information can be virtually unintelligible. To make matter worse, the symbols that are used for pneumatics are generally the same symbols that are used for hydraulics, and each company usually has small variations in their symbology. The untrained engineer or mechanic will take one look at a piping, instrumentation, and distribution (PID) drawing and exclaim, "What the heck is this?"

For these reasons, many companies adopt a nonstandard, but intuitively understandable, set of pneumatic symbols. The symbol system is generally based on circles with a one- or two-letter code followed by a number. The PID will have a legend, which identifies the component. The identification will generally have a description, part number, and, in some instances, a reference to the vendor. (See Fig. 8-10.) Table 8-2 shows some of the more common "in-house" symbols that may be used. If a special component is used, it is drawn in the same fashion as the rest and assigned a special code.

To further illustrate the difference between the use of standard and in-house pneumatic symbols, let's take a look at a small compressed air apparatus. Figure 8-8 shows a typical cylinder control system. This arrangement may be found on all manor of industrial equipment and is a good arrangement to illustrate the difference between the use of standard and in-house pneumatic symbols. The assembly consists of an air feed terminated with a pressure gauge, ball valve, coalescing filter, particulate filter, and regulator. The termination feeds a four-way solenoid valve, which is equipped with a muffler, two speed control valves, and a cylinder. Figure 8-9 shows how this assembly would be represented using standard pneumatic symbols.

In contrast to standard symbols, Fig. 8-10 shows how the assembly would be represented using in-house symbols. The diagram uses only two different symbols with letter codes. The codes are described in the legend. This method of drawing pneumatics produces drawings that are very easy to read and understand, even by untrained personnel. Unless the drawing represents a production component, the part numbers are generally the same number as the manufacturer's. A third column is oftentimes added that references the vendor for the specific component. A vendor reference helps speed the process of acquiring replacement parts in the future. Most personnel won't remember where the parts are from five or ten years down the road.

TABLE 8-2   In-House Pneumatic Symbols

Hose		Plug	
Flow		Quick Disconnect	
Bidirectional Flow		Crossing Lines	
Cylinder		Connected Lines	
Muffler	MF	Cooler	C
Two-Way Valve	V	Temperature Gauge	TG
Three-Way Valve	V	Filter	F
Four-Way Valve	V	Dryer	D
Solenoid Valve	SV	Line Trap	LT
Pilot Valve	PV	Drop Trap	DT
Pressure Gauge	PG	Lubricator	L
Air Motor	M	Automatic Drain	AD
Regulator	R	Pressure Switch	PS
Needle Valve	NV	Relief Valve	RV
Flow Meter	FM	Safety Release Valve	SR
Speed Control Valve	SC	Check Valve	CV
Shuttle Valve	SV	Relief Valve	RV

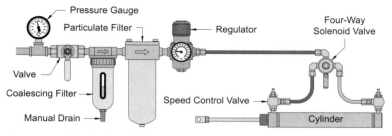

Figure 8-8 Cylinder installation.

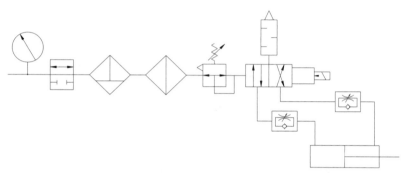

Figure 8-9 Schematic representation with standard symbols.

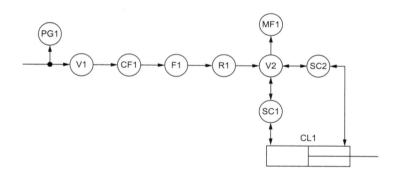

Code	Description	Part Number	Vendor Code
CF1	1/4" Coalescing Filter	216Z45-25	5
F1	1/4", 5-μm Filter	F-25-5-16	3
MF1	1/4" (M) NPT Muffler	22871	2
PG1	1/4" Bottom Mount, 200 psi, 2" Gauge	25-200-2	1
R1	1/4" Venting Regulator w/2" Gauge	221R13-25	5
SC1	1/8" Speed Control Valve × 90°	12-18-09	1
SC2	1/8" Speed Control Valve × 90°	12-18-09	1
V2	1/4" Four-Way Solenoid Valve, 115 VAC	6X247	2
CL1	1" × 12" Cylinder	B-001-12-C	4

Figure 8-10 Typical in-house schematic with legend.

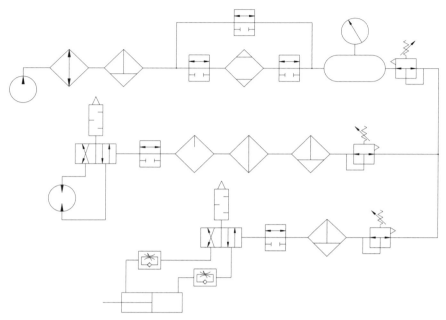

Figure 8-11 Schematic for Chap. 8 questions.

Questions*

1. What direction does air flow in the schematic shown in Fig. 8-11?

(A) Clockwise   (B) Counter clockwise   (C) Up   (D) Down

2. How many two-way valves are shown in the schematic?

(A) Seven   (B) Two   (C) Five   (D) Ten

3. What significant feature does the motor show?

(A) On/off operation   (B) Bidirectional   (C) Lubrication
(D) Slow RPM

4. How is the cylinder speed controlled?

(A) Four-way valve   (B) Restrictor orifice   (C) Speed control valve
(D) Ball valve

5. What two components is the lubricator between?

(A) Filter-valve   (B) Filter-trap   (C) Regulator-trap
(D) Dryer-regulator

_____

*Circle all that apply.

Electrical Controls

Most stationary compressors rely exclusively on electricity to power the pump and associated controls within the unit. For most of these units, turning on and off the motor is the primary duty of the unit's electrical controls. However, many compressors have a myriad of other electrical functions and controls. This chapter will provide you with a general understanding of the different systems that are commonly used. It should be noted that the examples shown are not reflective of any one compressor. For specific information on the controls of any given compressor, refer to the manufacturer's documentation supplied with that compressor.

Electrical controls on compressors can vary from a simple, single-pole pressure switch to extremely sophisticated microprocessor-based systems. A pressure switch control will simply turn the motor on and off in reference to the receiver pressure, while microprocessor-based systems might control energy consumption, flow rates, and a variety of system parameters. These systems can even control secondary compressors during peak and low flow periods.

Controls are usually housed in some sort of metal cabinet, which is mounted directly to the compressor. These cabinets should conform to NEMA (National Electric Manufacturers Association) standards. Generally speaking, lower-cost compressors will have their controls housed in a NEMA 3R enclosure, while larger and more expensive systems will typically use NEMA 4 or 4X enclosures. NEMA 3R enclosures are rated for indoor and outdoor use and are resistant to falling liquids and light splashing. NEMA 4 and 4X enclosures are rated for indoor and outdoor use, falling liquids, nonhazardous dust, lint and fibers, and wash-down applications. NEMA 4X enclosures are also rated for protection against corrosive agents. Obviously, a 4X enclosure is the preferred choice. I have opened 4X control cabinets on compressors that are thick with grease and dirt, only to be surprised with controls that look virtually brand new. On the other hand, old controls that are housed in a 3R cabinet are rarely salvageable. On one occasion, I opened a 3R enclosure on an outdoor compressor and was assaulted by about 35 geckos fleeing for their lives. Always be certain that the enclosure's door is closed and secured, usually with a single screw on 3R enclosures and two or more clamps or screws on 4 and 4X enclosures. A properly secured enclosure will keep out unwanted dirt and debris and will provide necessary protection against electrical shock. On another occasion when I was called out to service a compressor, I discovered that the enclosure door was standing open, nestled in the bottom of the cabinet was a bird's nest. A clear indication that the compressor wasn't receiving the routine maintenance that it required.

The following pages will take you through a basic education on the electrical controls that are used on most compressors. Take the time to carefully read and understand this chapter. This information will definitely payoff in the long run.

Most electric compressors operate on standard AC voltages. Smaller units, generally less than 2 hp, are normally 120 VAC, single phase. Units from 3 to 5 hp are normally connected to 240 VAC, single phase. Compressors in the 5- to 30-hp range generally operate on 220 VAC, three-phase power. Compressors over 30 hp will normally operate on 480 VAC, three-phase power. Larger compressors are generally set up to operate on high-voltage three-phase power because much lower currents can be utilized. This means that wire sizes can be smaller and, therefore, less costly.

All AC power in the United States is 60 cycles per second (Hz). What this means is that the polarity of the power reverses 60 times every second. Figure 9-1 shows a graphic representation of 60 Hz power at the three most common voltages. Take notice that a single cycle is the same

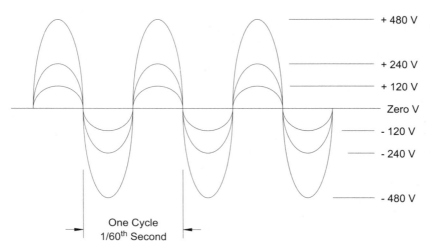

Figure 9-1   120, 240, 480 VAC 60 Hz Graphic.

($^1/_{60}$ of a second) for all three voltages. The additional power is derived in the amplitude or peak of the voltage. Approximately twice the total power of 120 VAC will be delivered by 240 VAC and 480 VAC will deliver approximately four times the power at the same current. This power is generally described in watts or, in the case of motors, kilowatts (kW) (1 W × 1000 = 1 kW). Amperage (amps) draw of a single-phase motor can be calculated by dividing total watts by the voltage (1000 W [1 kW] ÷ 220 VAC = 4.6 A). To determine the amperage per phase of a three-phase motor, divide the amps by 3 (4.5 A ÷ 3 = 1.5 A per phase).

Single-Pole Control

Compressors in the $^1/_2$- to 2-hp range generally operate on 115 VAC. The control system is usually a simple, single-pole pressure switch with a manual override. The switch is generally supplied with an integral electrical enclosure as well. Figure 9-2 shows a schematic for a small compressor. Figure 9-3 shows how a typical compressor would appear with a control system of this nature. Note that the manual on/off lever is an integral part of the pressure switch.

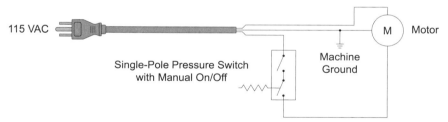

Figure 9-2   Single-pole control circuit.

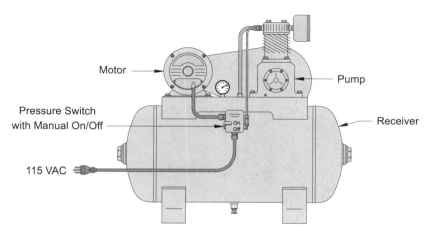

Figure 9-3   Compressor with manual/automatic pressure switch.

Double-Pole Control

Compressors in the 2- to 5-hp range generally operate on 240 VAC, single phase. The control system is similar to the 115-VAC system, except the pressure switch is a double-pole unit. These pressure switches usually do not incorporate a manual override lever, however, some units will provide this function. For units that do not have an override lever, the on/off function is generally provided by the power disconnect. As with the 115-VAC units, the switch is generally supplied with an integral electrical enclosure as well. Figure 9-4 shows a schematic representation of a 220 VAC, double-pole control circuit. Figure 9-5 shows how a typical

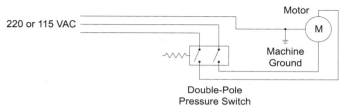

Figure 9-4   Double-pole control circuit.

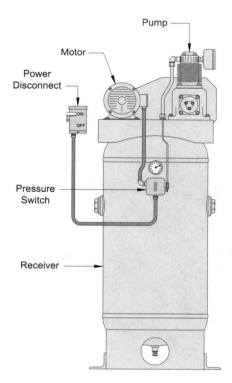

Figure 9-5 Compressor with double-pole pressure switch.

vertical compressor would appear with a double-pole control system. Note that the manual on/off is provided by the power disconnect.

Motor Starters

Compressors larger than 5 hp will typically be equipped with a motor starter. Because the pressure switch can't handle the high starting loads associated with larger motors, a motor controller is necessary. A motor starter consists of a set of contacts that are actuated by a magnetic coil and a motor overload. The pressure switch turns the coil on and off which, in turn, opens or closes the contacts of the controller. In this way a relatively inexpensive, single-pole pressure switch can control high horsepower motors. In addition to the contact set, motor starters have a set of "heaters" which will open the contacts if they see an overcurrent versus time condition. Manual on/off function may be built into the motor starter, but is usually provided by the power disconnect. Figure 9-6 shows a schematic representation of a typical three-phase motor controller with a pressure switch. Figure 9-7 shows how a commercial

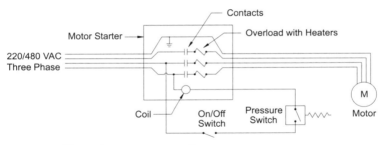

Figure 9-6   Three-phase motor controller.

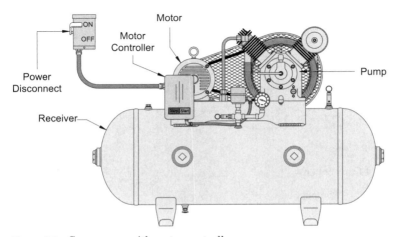

Figure 9-7   Compressor with motor controller.

reciprocating compressor might be set up with a three-phase motor controller.

120-VAC Control Loops

Lower-cost compressors, which are equipped with motor controllers, generally use the line voltage to actuate the coil. Compressors that target industrial applications will normally be equipped with a low-voltage control circuit, as shown in Fig. 9-8. The controller housing will have a control transformer, which accepts line voltage and produces 24 or 120 VAC. The control transformer should be equipped with two input fuses and one output fuse. This fusing arrangement will protect both the input lines and the output of the transformer from possible damage caused by an overload condition.

There are two distinct advantages of a system that operate with a control voltage. The first is that the compressor can be set up to use either 240 or 480 VAC. Motors from 5 to 50 hp are generally dual-voltage units and the control transformers will have an input that can be configured to accept either voltage. This attribute provides a certain amount of flexibility when installing a compressor. The second advantage is one of safety. All the control elements, including the on/off switch and the pressure switch, operate at nonlethal voltages. At the lower voltages, if there is a malfunction or ground fault within a component that an operator may come in contact with, it is unlikely that a serious injury will result.

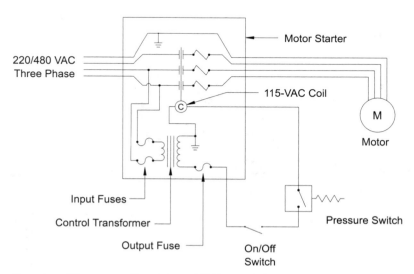

Figure 9-8   Motor controller with 120-VAC control circuit.

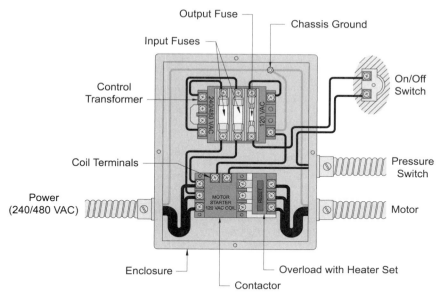

Figure 9-9 Commercial motor controller with 120-VAC control circuit.

Figure 9-8 shows a schematic representation of a typical motor controller with a 120-VAC control loop.

Figure 9-9 shows a typical motor controller layout with a 120-VAC control transformer. Many standard transformers have the input and output fuses mounted directly to the top of the transformer. Other systems may have the fuse set mounted to the cabinet adjacent to the transformer, or may have panel fuses. Fuse ratings should be clearly printed on the inside of the cabinet. If fuses are blowing often, do not use larger fuses. This is an indication that there is a problem within the control loop and proper repairs should be carried out by a qualified technician.

Sensor Loops

Figure 9-10 shows a motor controller schematic which incorporates a sensor loop. The sensors are used to monitor various parameters within the compressor. Typically output pressure, output temperature, oil pressure, and oil level will be monitored. Oftentimes, an emergency stop button will be placed into this loop. If any of the sensors sees a condition that is out of range, a switch will open and the control voltage will be interrupted to the contactor coil. For systems that have a low oil pressure sensor, a momentary start button must be added to the loop. Because the oil pressure switch is normally open, the coil will not be energized when the run switch is turned on. To start the compressor, the

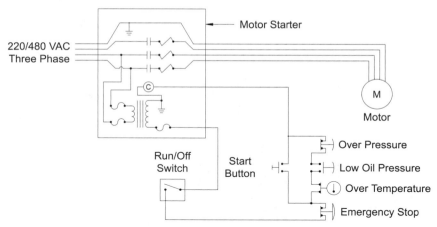

Figure 9-10   Motor controller with sensor loop.

start button is depressed and temporarily bypasses the sensor loop. After the oil pressure has built up to the operating range, the start button can be released and the compressor will run normally. This type of control circuit will usually be used on continuously run compressors.

Figure 9-11 shows a typical motor controller layout with a 120-VAC sensor loop. Take notice that the basic motor controller is of the same

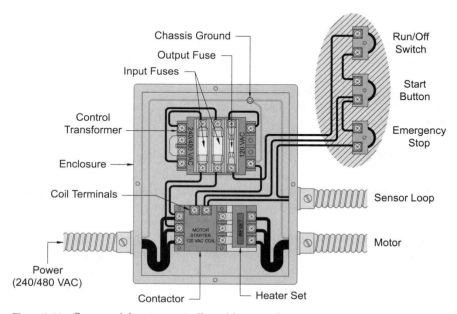

Figure 9-11   Commercial motor controller with sensor loop.

arrangement as the controller shown in Fig. 9-9. The principal difference is the addition of a start button, an emergency stop button, and the
various sensors, which are not shown. This is a very common control
system and is often found on smaller continuous-run screw compressors.

On, Off, and Automatic Functions

In many situations, compressed air installations are required to deliver
air over a varying range. The most common situation is a plant, which
may operate at 100 percent during the day, a substantially lower rate
during night shifts and at an even lower rate during weekends. To
accommodate these applications many manufacturers equip their screw
compressors with on, off, and auto functions. When the compressor is
switched to on, it runs continuously. When switched to automatic mode,
it will operate in the same fashion as a reciprocating compressor. When
the receiver reaches an upper limit, the compressor turns off. When the
receiver pressure drops to a predetermined lower limit, the compressor
turns on.

Figure 9-12 shows a schematic representation of a controller with on,
off, and automatic functions. The on/off/auto switch is a two-pole double

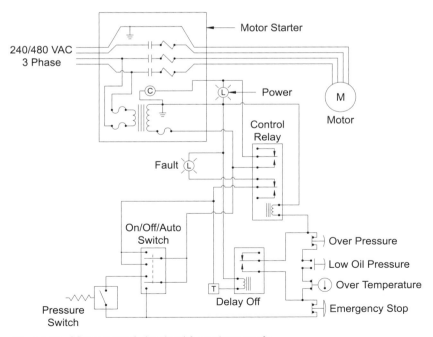

Figure 9-12   Motor control circuit with run/auto mode.

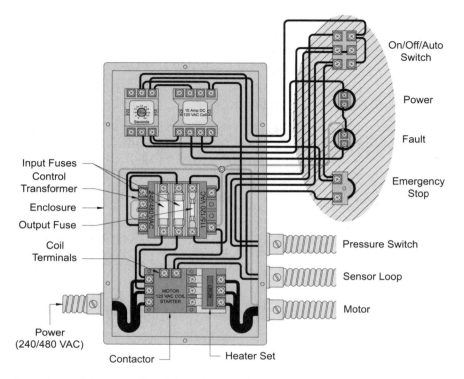

Figure 9-13   Motor controller with run/auto mode.

throw unit. The start button shown in Fig. 9-10 is replaced with a delay-off relay. When the compressor is turned on, the relay remains closed for a predetermined time to allow the oil pressure to come up to operating level. There is typically a control relay in these circuits, which controls the contactor and provides a fault light. If the sensor loop is interrupted, then the control relay disconnects the contactor coil and illuminates a fault lamp. To reset the system, the on/off/auto switch must be set to "off" and then back to "on" or "auto."

Figure 9-13 shows a typical on-off, auto controller. Note the two relays at the top of the enclosure. Also note the increasing complexity of the system.

Sensors

The sensors that are incorporated on air compressors come in two basic categories, the first being switches. Switches are simply an on/off switch coupled to some sort of mechanical actuator. Oil level switches, for example, are a housing with an internal float switch. The float is placed in a position that will open if the oil level drops below a predetermined level.

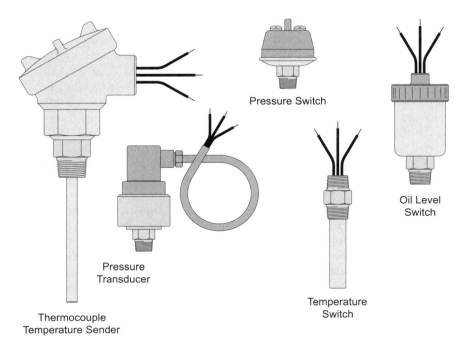

Pressure Switch

Pressure
Transducer

Oil Level
Switch

Thermocouple
Temperature Sender

Temperature
Switch

Figure 9-14 Various sensors.

A pressure switch will snap open or close when the system pressure reaches a predetermined limit.

The second type of sensors are transducers. These sensors transmit a range of information. As an example, a pressure transducer may have a range from 0 psi to 200 psi. The output of the transducer is generally a 0- to 10-V signal; 100 psi will equal 5 V, 150 psi will equal 7.5 V, and the like. Temperature transducers normally incorporate a thermocouple element. The output of a thermocouple requires a special receiver, either a stand-alone unit or a controller with a built-in receiver.

Figure 9-14 shows a few of the more common sensors that might be encountered on a typical air compressor.

Delta/Wye Motor Starters

One type of motor starter found on some compressors is the Delta/Wye starter. This type of starter is designed to start in a reduced power mode and is switched into a high-power mode for running. During start up, an electric motor may pull as much as seven times the current as it does when running. By starting the motor in a Wye configuration, the current, and consequently the torque, is about 40 percent less than starting the motor in a Delta configuration. Once the motor has come up to full speed, the starter switches the motor to a Delta configuration.

Normally, Delta/Wye starters are used on equipment with high inertial loads, such as punch presses, mill rollers, and press brakes. However, some companies are offering these types of starters as a cost-saving option on new compressors. The failure of this concept is that compressors do not have high inertial starting loads. Further, compressors much over 30 hp usually run continuously and are started, at most, once a day and oftentimes are started only after routine maintenance, once in every three months or so. With such infrequent starting cycles, the Delta/Wye starter doesn't produce any measurable savings whatsoever. Smaller reciprocating compressors that start and stop many times a day are simply not large enough to produce any significant energy savings from a starter like this. To make matters worse, the starter is very complex when compared to conventional starters. In the event that the starter fails, an outside technician usually has to be called in to conduct repairs, all the while your compressed air system is down. Figure 9-15 shows a schematic representation of a Delta/Wye motor starter.

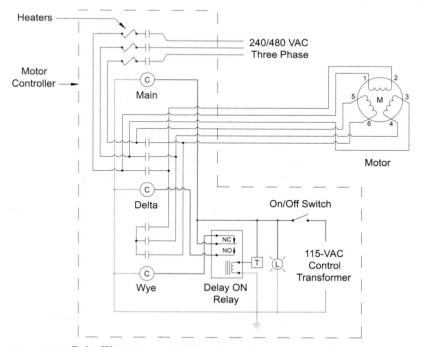

Figure 9-15   Delta/Wye motor controller schematic.

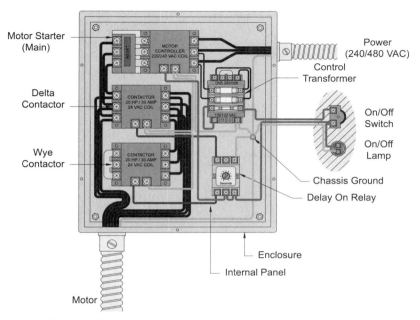

Motor Starter
(Main)

Delta
Contactor

Wye
Contactor

Power
(240/480 VAC)

Control
Transformer

On/Off
Switch

On/Off
Lamp

Chassis Ground

Delay On Relay

Enclosure

Internal Panel

Motor

Figure 9-16 Delta/Wye motor controller.

Figure 9-16 shows a typical Delta/Wye motor starter. Notice that there are three different contactors. The top contactor is the main unit and also carries the heater set. The middle contactor is the Delta contactor and the bottom the Wye.

Soft Starters

Soft starters are intended to provide the same basic function as a Delta/Wye starter, that is, reducing starting loads. These devices accomplish this by controlling a set of silicon-controlled rectifiers (SCR). SCRs are solid-state switches that may be controlled in a very precise manner. When turned on, these starters only allow a small portion of the electrical signal through to the motor. As time progresses, more and more of the signal is allowed to pass until the motor is ramped up to power.

Figure 9-17 shows a typical commercial soft starter. These starters generally have an on/off switch, a bypass switch, and start time control. The start time control allows the operator to select a ramp up time from 0 to 30 seconds.

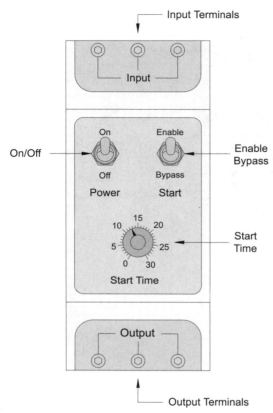

Figure 9-17 Commercial soft starter.

Figure 9-18 is a graphical representation of the ramp up cycle of a typical soft starter. The area in the dotted line is the delay time and the area in the solid line is the on time. Each time the signal passes the zero crossing, the starter is reset and the delay/on cycle starts over. Notice that the delay time gets progressively shorter and the on time gets progressively longer during the ramp up cycle.

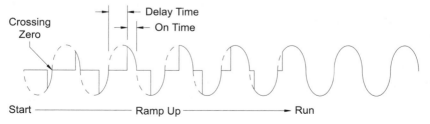

Figure 9-18 Switching cycle.

Variable Frequency Drives

One of the latest innovations in compressor control system is the variable frequency drive. These devices are AC motor speed controllers. An AC motor's speed is directly dependent on the frequency of the power line (60 Hz). If the power line frequency drops, the motor speed lowers; similarly, if the frequency increases, the motor speed increases. Variable frequency drives take advantage of this attribute.

The drive receives AC line voltage at 60 Hz and converts it into DC. The drives then syntheses a variable AC output, which is connected to the motor. When the drive is producing a 30 Hz output, the motor will operate at half speed. At 60 Hz, the motor operates at full speed and at 120 Hz the motor operates at double speed.

The biggest advantage of a variable frequency drive is that the motor produces its full torque even at lower speeds. This provides a drive that is ideal for manipulating larger compressors. If the compressed air load is at 60 percent then the motor drive can adjust the motor speed to 60 percent. This will provide a substantial energy savings over the operational life of the compressor. What used to be done by choking the input of the compressor can now be done by lowering the speed and, consequently, the operating cost.

Figure 9-19 shows a typical variable frequency drive.

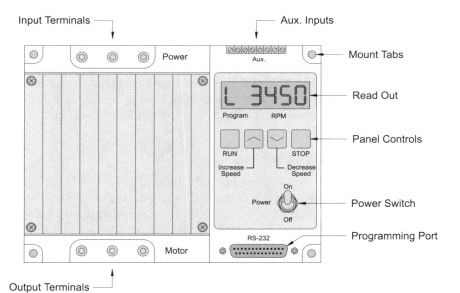

Figure 9-19   Commercial variable frequency drive.

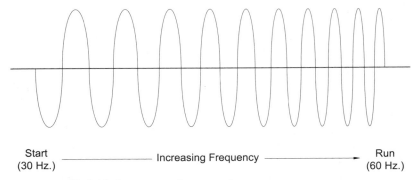

| Start
(30 Hz.) | ——————— Increasing Frequency ———————▶ | Run
(60 Hz.) |

Figure 9-20   Variable frequency soft start cycle.

Many companies are now offering variable frequency drive retrofits for existing compressors. Large screw, rotary vane, and turbo compressors are particularly suitable applications for a drive of this nature.

Most variable frequency drives are fully programmable. They are typically equipped with a front panel, which allows the operator to program the unit directly. In addition to the front panel, most drives are equipped with a serial port so that the unit can communicate with a host computer. The host computer can be used to monitor the various parameters of the drive or to adjust its program from a remote location. These types of drives will also have a number of auxiliary inputs that may be used to monitor the outputs of various sensors throughout the compressor. The auxiliary inputs are generally set up to accept 0 to 10 VDC signals.

Figure 9-20 shows a graphical representation of the output signal of a variable frequency drive that has been programmed to have a soft start-up cycle. The signal starts at 30 Hz (half speed) and ramps up to full speed by systematically increasing the frequency until it reaches 60 Hz.

Programmable Logic Controllers

Over the past few decades, programmable logic controllers (PLC) have grown to dominate the electrical control industries. These devices are low-cost industrial computers, which are optimized specifically for electrical control applications. They will typically have a series of inputs, a series of outputs, and a serial port for programming and communications. These are exceptionally friendly devices to the electrical designer. The wiring design is very straightforward and will be devoid of any relay logic. The logic of the controls is in the form of software. If a change is desired then it is simply, a matter of editing the program.

PLCs do have one significant drawback. If the PLC fails, then it must be repaired or replaced by the manufacturer of the compressor. This is because a replacement PLC is useless if it doesn't have the associated program and the program is proprietary to the manufacturer. It is common practice for manufacturers to demand a significant replacement cost for these components. Despite an inflated cost, the customer has no other recourse and is forced to make the purchase. This situation also holds true for after market companies that specialize in retrofitting equipment with PLCs.

Figure 9-21 shows a typical PLC-based compressor control. Take notice that all the different components are wired to individual terminals.

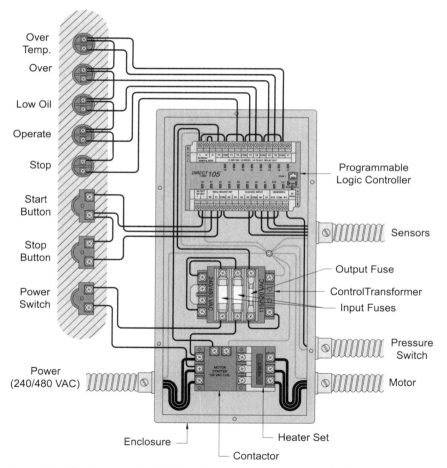

Figure 9-21 Controller with PLC. (*Courtesy of Automationdirect, Inc.*)

Microprocessor-Based Controllers

Most manufacturers now offer microprocessor-based controls on their compressors. These types of controllers are similar to a PLC-based system, except that they usually have some sort of operator programming panel. It has been my experience that a typical maintenance technician is generally unable to adjust the program parameters without special training. In some cases an operator trying to improve the operating parameters of a compressor will inadvertently shut down the entire compression system and have no idea what to do to get it back up.

On one occasion, I was called out to reprogram a compressor to adjust the output pressure. When I arrived on site, the compressors (two 250-hp units) were completely shut down. I was able to quickly reboot the systems, adjust the parameters, and restart the compressors. I found out later that a maintenance man had decided to lower the output pressure of the compressors, had read up on how to adjust the program to accomplish this and promptly rendered the compressors inoperable. This resulted in a 135,000 ft^2 manufacturing facility being completely shut down for 6 hours. When I arrived on site, everyone on the plant floor was simply standing around waiting for the air to come back.

Figure 9-22 shows a typical microprocessor-based compressor control. In this particular illustration, the motor controller is in the form of three SCRs. Motor protection is handled by the computer.

Figure 9-23 shows a microprocessor-based compressor controller with a variable frequency drive. A system like this represents cutting edge controls in the compression industry. Although purchase, repair, and replacement costs can be considerable, controls like this are extremely flexible and can be used to significantly reduce energy costs on almost any compressed air system.

Emergency Stop Circuit

Not all compressors have emergency stop circuits. Many of the compressor's emergency stop circuits simply interrupt the sensor loop. However, it is becoming increasingly common for compressors to have more traditional emergency stop circuits. These circuits are typically stand-alone arrangements that operate independently from the rest of the controls. Traditional emergency stop circuits are the preferred choice in this regard. These types of circuits can be identified by the panel controls. This type of design will have a large red palm button (emergency stop), a reset button, and a ready lamp. When the emergency stop button is depressed, it opens a relay, which disconnects the

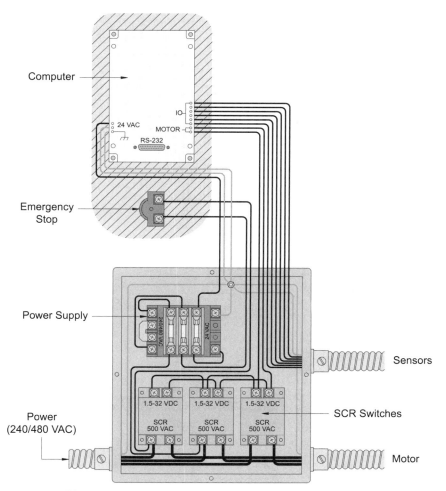

Figure 9-22 Microprocessor-based compressor controller.

contactor coil from the control transformer. At the same time, the relay also removes the emergency stop button so that the compressor will not restart when the button is reset. To reset the circuit, the emergency stop button is reset (if necessary) and the reset button is depressed. The ready lamp will illuminate and the compressor can be turned back on.

Figure 9-24 shows a schematic representation of a traditional emergency stop circuit. It should be noted that many manufacturers incorporate the ready lamp into the reset button.

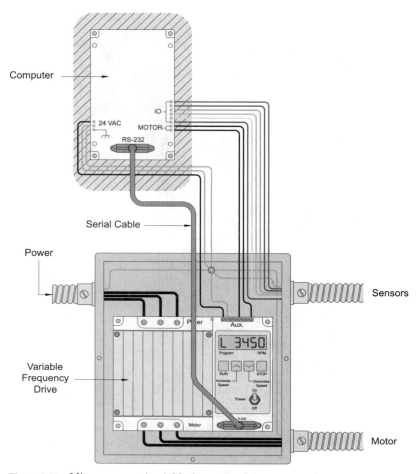

Figure 9-23 Microprocessor/variable frequency drive.

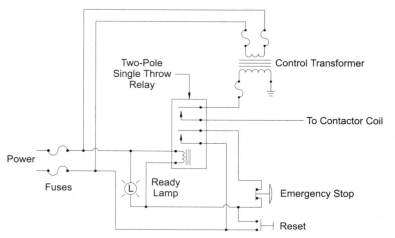

Figure 9-24 Emergency stop schematic.

Power Disconnects

Any electrically operated compressor that is permanently wired to the power source should incorporate a power disconnect which is mounted adjacent to the compressor. Power disconnects are generally housed in a NEMA 3R cabinet. The switch mechanism carries a dual action safety interlock that prevents the power from being turned on if the door is open and prevents the door from opening if the power is on. If for any reason this safety mechanism is nonfunctional, then the disconnect should be replaced. Most disconnects also carry a fuse set, which is normally in the form of standard cartridge fuses.

Figure 9-25 shows the internals of a commercially available power disconnect. Figure 9-26 shows the disconnect with the door closed.

Service Outlets

It's a good idea to add a 120-VAC service outlet adjacent to any power disconnect. A service outlet is an exceptionally handy thing to have access to when conducting maintenance to the compression system. If the disconnect is wired for 240 VAC or 480 VAC, then a separate 120 VAC line must be run, as shown in the illustration, or a step-down transformer can be installed. If a transformer is used, be certain that the unit is equipped with both input and output fuses. In either case, the service outlet should remain on even when the power disconnect is switched to off.

Figure 9-26 shows an example of a service outlet, with switch, mounted adjacent to the power disconnect.

Lockout/Tagout Programs

When maintaining any piece of equipment it is essential to turn off and lock out the power. Most companies have a regular lockout/tagout program that maintenance personnel are required to follow. If your company does not have a program in place, then one should be adopted. Locking out the compressor will prevent anyone from turning it on during maintenance operations. Maintenance may be taking place in an area of the compression system, which is out of sight of the compressor. If someone turns on the compressor, things can get very exciting in short order. If only one maintenance operation is being carried out, then a simple padlock can be placed on the power disconnect. If several maintenance operations are being carried out simultaneously, then a multi-lock hasp is incorporated. Each maintenance group should be assigned a color-coded padlock and the lead man of each group is the only person that should have a key to the group's assigned lock. Additionally, each lock used should have a clearly visible warning tag with contact information for the key holder.

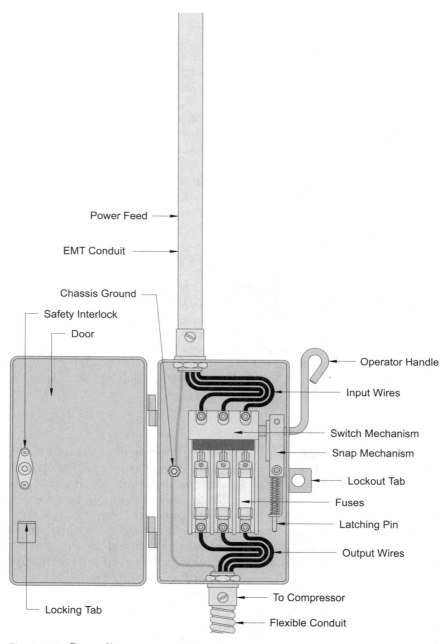

Figure 9-25   Power disconnect.

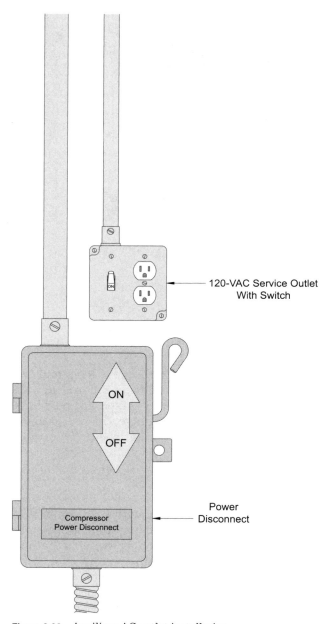

120-VAC Service Outlet
With Switch

ON

OFF

Power
Disconnect

Compressor
Power Disconnect

Figure 9-26   Auxiliary AC outlet installation.

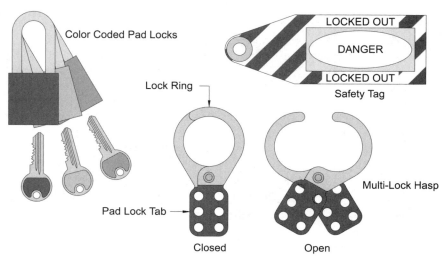

Figure 9-27   Lockout safety items.

Figure 9-27 shows some typical lockout tools. It should be noted that color-coded lockout padlocks are available from many industrial supply houses.

Figure 9-28 shows a power disconnect with a lockout hasp and two color-coded padlocks. Tags should be large enough to present a clear indication that the equipment is out of service. Many companies may have a workforce that is not fluent in the English language, this is especially true in Southern states. In these cases, lockout tags should be printed in both English and the secondary language.

Duplex Controllers

Most packaged duplex compressors are equipped with a controller that toggles the pumps during normal operation and operates both pumps during peak load periods. To achieve this function for systems that use two independent compressors, an external duplex controller must be added as shown in Fig. 9-29. The duplex controller is operated by the pressure switches that are supplied with the compressors. Pressure switch number 1 is used for normal operations and pressure switch number 2 is set up to trigger peak demand situations. These units generally have a bypass switch, which allows the compressors to operate in parallel.

A typical duplex controller is a relay logic circuit housed in a small NEMA cabinet. There are a number of different circuit designs on the

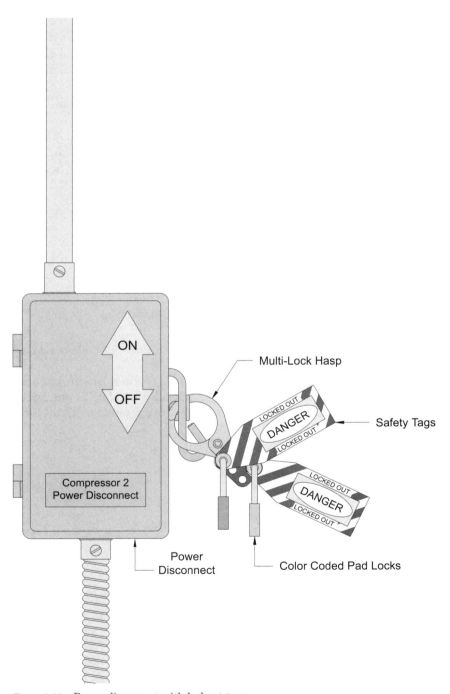

Figure 9-28   Power disconnect with lockout tags.

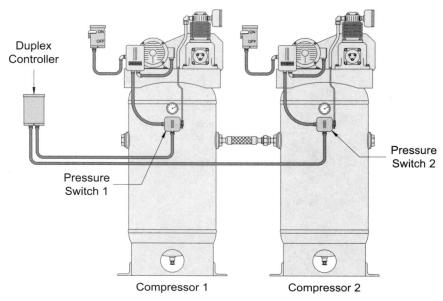

Figure 9-29   Two separate compressors set up with a duplex controller.

market, including some units that use PLCs. Figure 9-30 shows a typical relay circuit for a duplex controller.

The heart of the controller is a latching relay. When the unit is activated by pressure switch number 1, the latching relay is energized through the "delay on" relay. The "delay on" relay feeds a signal for a

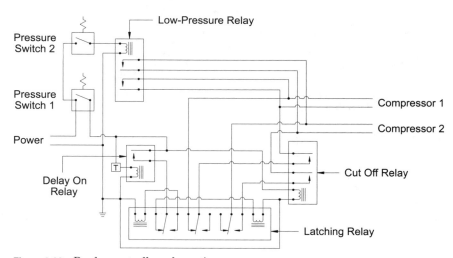

Figure 9-30   Duplex controller schematic.

period just long enough to engage the latching relay. At the same time, the "cut off" relay is energized and the compressor starts. When the system pressure is reached, pressure switch number 1 breaks the signal to the "cut off" relay and the compressor turns off.

During normal operation, if the system pressure drops below a pre-determined limit, pressure switch number 2 energizes the "low pressure" relay. This relay bypasses the entire latching circuit and turns on both compressors until the system pressure is reached.

Parallel Duplex Operation

For duplex installations under about 20 hp, a stand-alone duplex controller doesn't provide much of a benefit. They are expensive, and will add a considerable level of complexity to your compression system.

A better alternative for these smaller systems is parallel operation. This is simply setting up the compressors so that they turn on and off at the same time. To accomplish this, two criteria must be met. First, the compressors must operate with the same control voltage and second, their motor controllers must be interconnected. The interconnection entails disconnecting the number 2 pressure switch and connecting the coils in parallel. Figure 9-31 shows a schematic representation of two compressors set up for parallel operation. If so desired, switches can

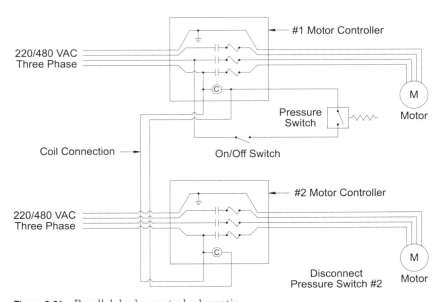

Figure 9-31 Parallel duplex control schematic.

be installed to isolate each coil so that either compressor can be taken off line for service without disturbing the operation of the other.

Weekly and Daily Duplex Operation

Larger compressors are usually set up for redundant operation, and it is a good idea to operate both compressors on a regular basis. Most large screw systems toggle their compressors on a 7-day cycle. That is to say, that one compressor operates for a week and the other operates for the next week. A 24 h/7 day timer can be set up to control installations such as this.

These duplex controllers generally have two timer modes, a 24-h toggle and a 7-day programmable timer. The 7-day timer can be set up to turn on and off the compressors in accordance with various days of the week as well as hours of the day. What this means is that the compression system may be set up to operate specifically in reference to the plant's operational hours. Over the long term, this can translate to considerable energy savings.

Figure 9-32 shows a typical 24 h/7 day duplex controller. Note that the controller has manual override switches for both compressors. This is to provide the necessary control so that routine service can be carried out without disturbing the operation of the compression system.

Figure 9-33 shows a schematic representation of the 24 h/7 day duplex controller. The core of the controller is the 7-day timer. This timer is set to correspond to standard time and runs continuously. A weekly pattern is programmed into the timer, and the compressors in the system are under 365-day-a-year control.

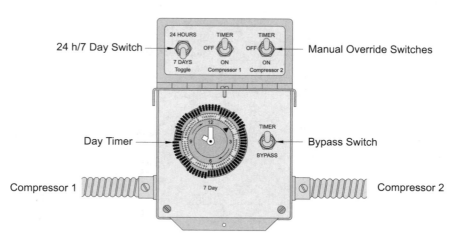

Figure 9-32   24 h/7 day toggle controller.

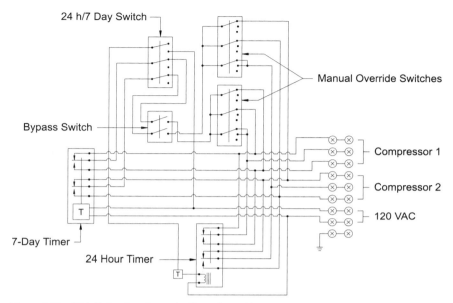

Figure 9-33 24 h/7 day toggle controller schematic.

Wire Types

When connecting a compressor, it is extremely important to consider what size and type of wire will be used. Generally, this decision should be the responsibility of a licensed electrician who has been brought in to install the electrical feed and interconnect the compressor. However, it is a good idea that anyone who has been assigned responsibility for overseeing the installation of a compression system should have a cursory understanding of this area.

The current that a compressor will require and the distance from the electrical source will determine the wire size necessary for any given installation. Figure 9-34 shows the relative sizes of standard American Wire Gauges (AWG). Generally EMT conduit is used to route the wires

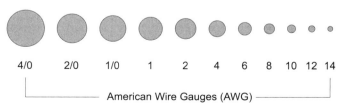

Figure 9-34 Standard wire sizes.

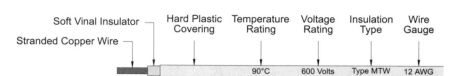

Figure 9-35   Type MTW wire.

to the compressor site, and wires are pulled through after installation. The wires should be stranded and have an insulation rating of at least 600 V. Additionally, the insulation type should be impervious to oil and water exposure. Generally, MTW-type insulation is specified for these applications. In any case, the wire selected should be clearly marked with the gauge, insulation type, voltage rating, and temperature rating, as shown in Fig. 9-35. Additionally, if the wire is to be pulled through conduit, then it should have a hard plastic covering over the primary insulation. The plastic cover will allow the wire to slide easily through the conduit.

Din Rails and Wire Guides

It is becoming increasingly common for manufacturers of electrical equipment to incorporate Din rails to mount components. The Din rail is a standard mounting system that greatly simplifies the construction of electrical controls. Figure 9-36 shows a length of Din rail and two snap-on relay sockets.

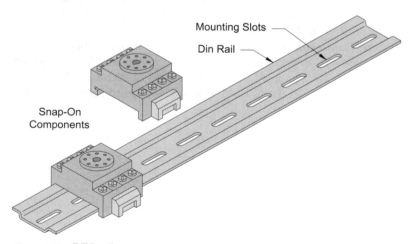

Figure 9-36   DIN rail.

Wire guides are another device that has become rather common. This component is a plastic tray that has a continuous array of vertical slots and a snap-on top. Wires can be routed through the tray and out any slot adjacent to the component that the wire must be connected to. After wiring is complete, the snap-on tops are installed and the entire control system takes on an extremely clean appearance. Figure 9-37 shows a typical wire guide.

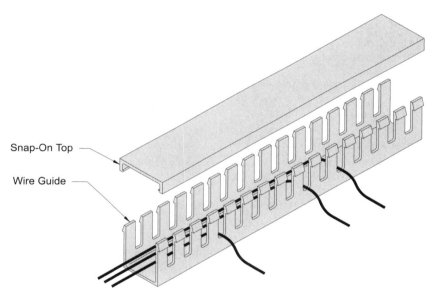

Snap-On Top

Wire Guide

Figure 9-37 Wire guide.

Questions*

1. What does NEMA stand for?

   (A) National Electric Motor Association
   (B) Normal Electric Manufacturing Applications
   (C) National Electrical Manufacturers Association
   (D) Nuclear Electric Standards and Motor Authority

2. A NEMA 4 cabinet is resistant to:

   (A) Water   (B) Oil   (C) Acid   (D) Dirt

3. What function does the motor controller serve?

   (A) Belt tension   (B) Pressure regulation   (C) Cooling
   (D) Motor control

4. What function do the heaters serve?

   (A) Prevent freezing   (B) Warm-up   (C) Motor protection
   (D) Comfort

5. Why is a low control voltage used?

   (A) Safety   (B) Smaller wires   (C) To add expense
   (D) To operate lamps

6. What type of sensors are commonly found on compressors?

   (A) Low oil   (B) Air pressure   (C) Temperature   (D) Oil pressure

7. What device should be used when connecting the power to a compressor?

   (A) Fan   (B) Wire nut   (C) Power disconnect   (D) AC outlet

8. What procedure is critical when servicing a compressor?

   (A) Oiling   (B) Cooling   (C) Cleaning   (D) Lock out

9. What type of duplex controller is appropriate for redundant screw compressors?

   (A) Peak demand   (B) 7 Day/24 h   (C) Electric   (D) Pneumatic

---

*Circle all that apply.

10

Maintenance

As with any complex system, compression systems benefit from regular, scheduled maintenance. Unfortunately, many compressed air installations receive rather poor maintenance, generally in the form of unscheduled repairs that are required only when the compression system has failed entirely. The plant shuts down and there is a frantic effort to get the system backup in any way possible. This method usually leads to substandard and inefficient repairs, which become the status quo. Compression systems that receive this type of maintenance are normally rife with leaks, unsound plumbing, and dangerous applications, which, in the long haul, costs the company substantial amounts of money. Another common maintenance problem is allowing the general employees to conduct maintenance duties. These employees rarely have adequate training for these duties and most of the time, efforts are carried out as an afterthought. After all, these employees have other, more important, duties that normally take priority over the compression system.

The best way to ensure that your compression system receives the scheduled maintenance that it requires is to develop a standardized system maintenance program. The program should identify service and inspection intervals for every item within the system. This may sound like an overwhelming task, but if you break it down into sections, assign certain duties to properly trained personnel, and set up a maintenance calendar, it turns out to be rather easy. All major services that require venting the system should be scheduled during the times when the plant is shut down, or at least minimized, such as weekends, evenings, or during third shift. Systems that have redundant compressors can have most of their major service done without disturbing the delivery of compressed air to the plant. Similarly, filters, lubricators, and

Figure 10-1   Valve/universal hose coupling.

applications that have isolation or bypass valves can be serviced without interrupting the plant's function.

Compression systems that have had many years of haphazard maintenance may have a large number of serious problems that need to be addressed. In these cases it can be difficult to clearly identify and correct all the existing problems. If you find yourself in a position like this, it is best to conduct a thorough air audit on the system. After the air audit is complete, systematically correct all the identified problems. On completion of repairs, you'll be dealing with a system that is in good, serviceable condition, at which point a preventive maintenance program can be designed. For more information on conducting an air audit, refer to Chap. 14. Figure 10-1 shows a gate valve equipped with a universial hose coupling.

Supply Side

In dealing with the supply side, the first significant piece of equipment that should be addressed is the compressor. The compressor should be maintained in strict accordance with the manufacturer's recommendations. Compressor manufacturers always supply recommended maintenance schedules with new units. If you have an old unit or are purchasing a used unit, most manufacturers can supply you with the original maintenance schedules for your particular compressor. In any case, the following items should be regularly addressed in any maintenance program:

1. Oil change
2. Oil filter
3. Air filter
4. Belt tension
5. Motor lubrication
6. Receiver drain and
7. Aftercooler cleaning

Other items that should be subject to regular inspections, but at a lesser interval, are:

1. Oil temperature
2. Output temperature
3. Duty cycle (for compressors that cycle)
4. Cut-in and cut-out settings
5. Motor temperature
6. Pressure switch contacts
7. Motor controller contacts
8. Safety valves
9. Pressure gauge
10. Check valve
11. Automatic drain settings and functionality
12. Cleanliness and
13. General condition such as fastener tightness and electrical terminations

Before shutting down the compressor for service, isolate it from the distribution system by closing the output valve. Turn off the power to the compressor. Be certain that a lockout tag is applied to the power disconnect box in accordance with OSHA regulations. It is also a good idea to place a lockout tag on the output valve. The next step is to vent all the pressure from the receiver. The safety relief valve or the drain valve can be utilized for this purpose. When using any valve to vent a large quantity of compressed air there exists the possibility of freezing. The water in the compressed air freezes inside the bore of the valve because of the expansion of the air. This can clog the valve and it may seem like the receiver is fully vented, but *it's not*! Always use the pressure gauge to determine the pressure of the receiver. In the event that the pressure gauge is nonfunctional, the pressure status can be verified by pulling the manual override ring on the safety valve. The final step is to allow the compressor to cooldown to room temperature. Compressors operate at elevated temperatures, which can easily cause severe burns. It is not uncommon for a reciprocating compressor to have surfaces that exceed 350°F! After cooldown, the compressor is ready for service. Figure 10-2 shows some of the areas on a typical packaged reciprocating compressor that should be inspected or serviced on a routine basis.

If the compression system has a stand-alone aftercooler, it should be cleaned and inspected on a regular basis. An aftercooler that is dirty is considerably less efficient than a clean one.

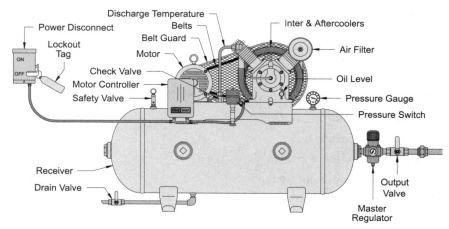

Figure 10-2 Maintenance items.

The next significant item to maintain is the refrigerated dryer. On CFC dryers, the condenser coil requires cleaning on a regular basis. It is generally a good idea to clean the coil at the same interval as the after-cooler. If the unit is equipped with a gauge set, verify that the input temperature is at or below the unit's maximum rating. The refrigeration gauges should operate in the green zones or within the range specified by the manufacturer. If the unit is not equipped with a gauge set, a licensed CFC technician should be called in to verify the refrigeration system. The water discharge rate on Elliott cycle dryers should be verified at least once a year. These types of dryers also benefit from a yearly decalcification and general inspection to ensure that the control system is functioning properly.

If the system is equipped with a twin tower desiccant dryer, the unit should undergo a regular general inspection. The output dew point should be verified on a scheduled basis. If the dew point rises above an acceptable level, then the desiccant should be replaced.

The receiver is an area, which is oftentimes neglected. The entire receiver should be inspected for rust spots. It is critical that any rust spots are addressed immediately. These spots can rust through the wall of the receiver, at which point the only recourse is to replace the tank—an expensive proposition when you have a 2000-gal receiver. The operation and settings for the automatic drain should be verified. I have encountered receivers that are actually filled with water because the automatic drain failed and no one ever bothered to check it. The first indication that the receiver is filled is when your applications start receiving straight water instead of compressed air.

The oil/water separator is another neglected area. If the system has one at all, they are usually ignored and are well past required maintenance. The discharge filter elements should be replaced and the separated oil level should be checked and drained as necessary.

Finally, the master regulator should be subjected to a general inspection. It should be clean and the output gauge should be in good serviceable condition. Check that the output is set at 90 psi. This setting has a tendency to be changed by employees who want a little more power from their tools. The output setting should never exceed 90 psi unless a special pressure is required.

It's a good idea to maintain good general housekeeping around the compressor. Compressors are usually treated as the dirty little secret in the back of the building and as such people have a tendency to dump all sorts of things in that area. This should be aggressively discouraged. Piling junk around the compressor restricts ventilation as well as access, it creates a safety hazard and is generally unsightly. The ambient conditions should also be considered. An enclosed compressor room in the south can easily have an inside temperature of 130°F! High ambient temperatures will add a significantly higher water load and have a negative effect on the unit's efficiency. Compressor rooms should be well ventilated, especially during the summer months. Inversely, all compression equipment should be protected from freezing temperatures during winter months. If the equipment is exposed to freezing temperatures, then water within delicate components may freeze and the components may malfunction or be ruined completely. For convection ventilation, large vents should be installed at the bottom and the top of the compressor room on at least two sides. If venting the walls is not practical, then a forced ventilation system should be installed. In areas, which have both high- summer temperatures and low winter temperatures, the compressor room should be equipped with removable panels that can be opened for warm weather and closed for cold weather. Figure 10-3 can be used as a guide for ventilating a typical compressor room.

Distribution

All air compressors require some sort of distribution system. This may be as simple as a regulator with a quick disconnect on a contractor's compressor or as complex as an instrument air system supplying a 200-acre petrochemical plant. Although the distribution system generally represents the most reliable portion of a compression system, it will require a certain amount of regular maintenance. The entire piping system should be inspected for leaks on a regular basis. Leaks do crop up, especially around valve stems, filters, regulators, lubricators, and disconnects. Pay particular attention to universal hose couplings and industrial

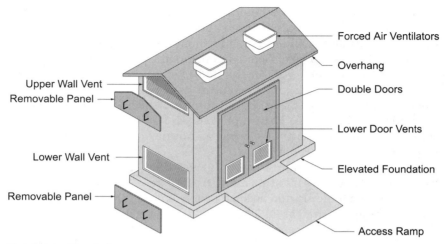

Figure 10-3 Stand-alone compressor room.

quick disconnects. These devices have a tendency to rapidly develop leaks. Universal hose couplings can usually be repaired by replacing the gaskets. Industrial quick disconnects should be replaced and the old unit discarded. Filter elements should be replaced according to a defined schedule. Lubricator oil levels need to be checked and topped off frequently. Regulators should be inspected and their output pressures verified. Be aware that employees have a tendency to increase output pressures when anything is not working properly at their workstation. All the system traps should be drained once daily, and more often if necessary. If line dryers are used, these units should receive the same basic inspections and maintenance as the system dryers. Figure 10-4 shows a gate valve with bolt together flanges.

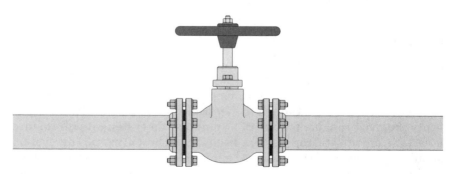

Figure 10-4 Bolted gate valve.

Applications

Applications are defined as: any use of compressed air after the distribution connects, including air hoses. Air hoses are oftentimes neglected because they are generally moved about the plant and are rarely returned to a central location. Hoses should be issued and returned to the tool crib in the same manner as any other tool. All hoses should carry a label, which identifies the particular assembly, its date of introduction and any pertinent maintenance information. The tool crib and labeling provide a perfect arrangement for conducting periodic inspections and maintenance. A cabinet and workbench should be set up adjacent to an air source and mop sink. The cabinet should be stocked with all the necessary components to conduct repairs to the hoses. The air source and mop sink are used to inspect for leaks and to carry out general cleaning.

Air tools should be inspected on a regular basis, whether they belong to the company or the employee. The employer should offer free inspection and service for employee-owned air tools. Oftentimes employees are unwilling to spend the money required to have routine inspections and service carried out on their personal tools. If an employee-owned tool is operating in a substandard manner, production suffers, and that costs the company money, not the employee. For this reason, many companies do not allow employees to have personal air tools, they are required to use company-issued tools. In any case, the tools should be in good, serviceable condition, with all safety guards in place. There should be no detectable leaks when the tool is at rest. Verify that the tool is receiving proper lubrication. If a lubricator is not used in the distribution system, then the tool should be lubricated every day of operation in accordance with the manufacturer's recommendations. The functionality and performance of the tool should also be verified in accordance with the manufacturer's recommendations.

Pneumatic equipment and controls can represent rather complex systems by themselves. Each piece of equipment should undergo regular inspections and service in strict accordance with the manufacturer's recommendation. Leaks, especially those that are imbedded deep within a mechanism, can be difficult to detect and great care should be exercised when inspecting complex pneumatic equipment. Benchmarking performance can be another effective way to gauge the condition of a piece of equipment. If the equipment is operating at a substantially slower rate than it did 6 months earlier, it's a safe bet that there is a problem that needs to be addressed. This could be the result of an internal leak, poor lubrication, seal deterioration, a bent piston rod, and the like.

Many pneumatic controls are designed in-house. The designers of these controls should supply maintenance personnel with a recommended inspection and maintenance schedule. The schedule can be

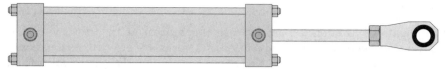

Figure 10-5 Bolted cylinder.

based on the recommendations from the manufacturers of the components used in the system. If there are no recommendations to work from, an aggressive schedule should be implemented and slowly relaxed as experience with the control system grows. As with any compressed air system, pneumatic controls benefit from proper pressure, flow, filtration, and lubrication. These items are firstly the responsibility of the designer and later the responsibility of the maintenance personnel. Closely adhering to the operational parameters and maintenance schedules of any control system, irrespective of whether it's on a stand-alone piece of equipment or a plant-wide system, will greatly limit unscheduled downtime and costly interruptions in production. Figure 10-5 shows a typical bolttogether air cylinder.

Maintenance Program

A comprehensive maintenance program should encompass every aspect of the compression system and all its applications. A list of all maintenance requirements should be made which includes time intervals, inspection criteria, expendables, settings, and personnel assignments. Individual maintenance requirements should be broken up and assigned to properly trained personnel. Weekly sign-off sheets should be made for each responsible person and these sheets should be filled out and filed after completion. The sheets should provide a check-off list, which indicates the date on which the assignment has been completed. In addition to the check-off list, the sheets should include a section for special notes, observations, or recommendations. These sheets will provide a running log of all maintenance activities in reference to the compression system. In some instances it may be beneficial to maintain a formal logbook with abbreviated references to the sign-off sheets. The sign-off sheets and logbook will aid in compiling maintenance reports, which are helpful in projecting future requirements.

When setting up a maintenance program, it is helpful to generate a calendar, which clearly identifies all scheduled maintenance duties. Large desktop calendar pads or one-year wall planners are ideal for generating these calendars. After a maintenance list is compiled, notations are made on the days, weeks, and months of the calendar. The calendar can be hung in a prominent location and it will guide personnel for the entire year. As each month is completed, the page can be torn off, folded, and filed. When filing maintenance records, it is important to know when enough is enough. Generally speaking, records over five years old are not much use. These records can be discarded or placed into long-term storage. Tables 10-1, 10-2, and 10-3 can be used as a

TABLE 10-1   Supply Maintenance Items

Supply	Daily	Weekly	Monthly	Six Month	Yearly	As Required	Per Manufacturers Recommendation
Check Compressor Oil		●					
Change Compressor Oil							●
Check Drive Belts			●				
Change Drive Belts						●	
Clean Air Intake Filter		●					
Change Air Intake Filter						●	
Lubricate Motor							●
Drain Receiver	●						
Clean Inter Cooler			●				
Clean Aftercooler			●				
Check Duty Cycle		●					
Verify Pressure Settings		●					
Inspect Pressure Switch Contacts				●			
Inspect Contactor Contacts				●			
Verify Safety Valve				●			
Verify Pressure Gauge				●			
Verify In-Tank Check Valve				●			
Clean Compressor				●			
General Inspection				●			
Verify Delivery Pressure (90 psi)		●					
Inspect Receiver				●			
Clean CFC Dryer Condenser			●				
Verify Dryer Readings			●				
Decalcify Elliott Cycle Dryer					●		
Verify Desiccant Performance				●			
Check Receiver Drain Settings			●				
Check Oil/Water Separator				●			
Change Oil/Water Separator Filter					●		
Check Compressor Room Temp.			●				
Clean Compressor Room			●				
Adjust Ventilation			●				

TABLE 10-2   Distribution Maintenance Items

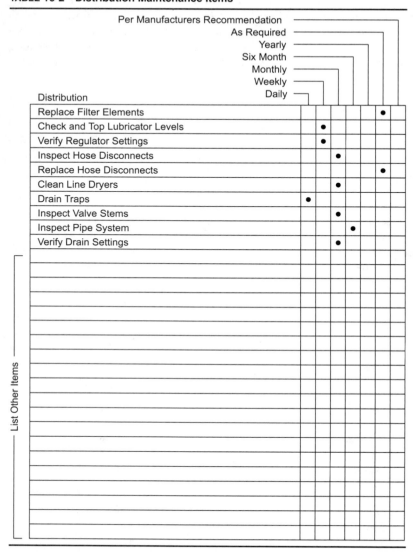

Distribution	Daily	Weekly	Monthly	Six Month	Yearly	As Required	Per Manufacturers Recommendation
Replace Filter Elements						•	
Check and Top Lubricator Levels		•					
Verify Regulator Settings		•					
Inspect Hose Disconnects			•				
Replace Hose Disconnects						•	
Clean Line Dryers			•				
Drain Traps	•						
Inspect Valve Stems			•				
Inspect Pipe System				•			
Verify Drain Settings			•				

List Other Items

guide to estimate service and inspection intervals for some of the more common components within a compression system.

Replacement Parts Inventory

There are a myriad of components and parts that require regular replacement on a compression system. In most cases, stocking a selection

TABLE 10-3   Application Maintenance Items

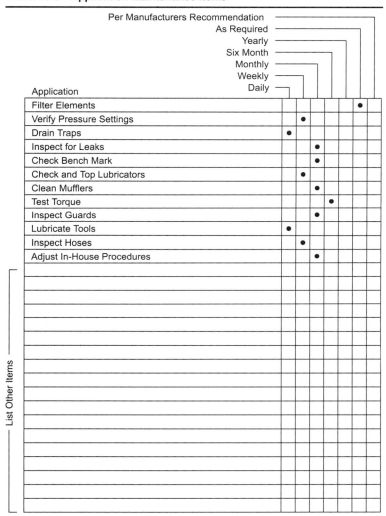

Application	Daily	Weekly	Monthly	Six Month	Yearly	As Required	Per Manufacturers Recommendation
Filter Elements							•
Verify Pressure Settings		•					
Drain Traps	•						
Inspect for Leaks			•				
Check Bench Mark			•				
Check and Top Lubricators		•					
Clean Mufflers			•				
Test Torque				•			
Inspect Guards			•				
Lubricate Tools	•						
Inspect Hoses		•					
Adjust In-House Procedures			•				

List Other Items

of replacement parts is less expensive than the downtime, which may occur if a small part fails during the course of production. Generally speaking, any reasonably inexpensive part that is critical for the continued operation of the system should be inventoried. "Reasonably inexpensive" is a subjective term and what that means will vary from installation to installation. As an example, to a production facility that may be worth a million dollars an hour, a $2500.00 part that has to be shipped in from out-of-town is reasonably inexpensive. To a four-bay auto repair shop that's more that they paid for the entire compressor. The replacement part inventory should match the specific requirements of the system that's being dealt with. The following list will give you some idea of replacement parts that are usually stocked.

1. Hose
2. Hose couplers
3. Hose clamps
4. Coupler gaskets
5. Compressor oil
6. Tool oil
7. Drive belts
8. Pipe fittings
9. Flange gaskets
10. Flange bolt sets
11. Fuses
12. Pipe
13. O-rings
14. Unloader valve
15. Pressure gauges
16. Regulators
17. Filter elements
18. Critical cylinders
19. Ball valves
20. Flex couplings
21. Low-cost air tools
22. Safety valves
23. Head check valves
24. Filter housings
25. Lubricators
26. Oil/water filters
27. Desiccant
28. CFC
29. Check valves
30. Drain valves

31. Strainers

32. Speed controllers

33. Solenoid valves

Questions*

1. What type of maintenance do compression systems benefit from?
(A) Random   (B) Regular   (C) Scheduled   (D) None

2. What is the best method for ensuring proper maintenance?
(A) Standardized system   (B) Regular checks   (C) Sound
(D) Time

3. Venting the system should be done at what time?
(A) Weekend   (B) Off-hours   (C) Third shift
(D) During minimized production

4. Name the three fundamental sections of a compression system.
(A) Application   (B) Distribution   (C) Drying   (D) Supply

5. The system dryer is located in what section?
(A) Distribution   (B) Drying   (C) Supply   (D) Application

6. What two items are important when inspecting the compressor?
(A) Desiccant   (B) Oil level   (C) Time of day   (D) Belt tension

7. When venting a receiver what phenomenon must you be aware of?
(A) Water mess   (B) Dirt   (C) Valve freezing   (D) Stink

8. How hot can the surfaces of a compressor pump get?
(A) 125°F   (B) Less than 200°F   (C) Greater than 350°F
(D) Around 185°F

9. What is the most important attribute of a compressor room?
(A) Housekeeping   (B) Cleanliness   (C) Ventilation
(D) Security

10. How often should an aftercooler be cleaned?
(A) Weekly   (B) Monthly   (C) Yearly   (D) As necessary

---

*Circle all that apply.

Energy and Costs Associated with Compressed Air

The energy costs associated with compressed air are rarely considered as part of the cost of manufacturing. However, these are very real expenses that always increase the general overhead of the business. In addition to the energy costs, there are a number of other expenses associated with the generation of compressed air. The energy and ancillary expenses of a compression system can be greatly minimized by simply being aware of all the pitfalls that may await you.

There are many areas within the compression system that can be fine-tuned in order to reduce operating costs. However, it is important to remember that the number one loss associated with compressed air is leaks. Some industry sources estimate that as much as 25 percent to 40 percent of all compressed air generated is lost to leaks. As an example of how much this can cost industry, take the case of a small manufacturing facility that operates 5 days a week with two 8-h shifts a day. They operate a 125-hp compressor that is leaking 17 percent of the air being generated. This leak rate costs the company over $2167.50 a year! Enough money to pay for a rather nice company picnic.

The second biggest loss associated with compressed air is misuse. Misuse varies from the blatantly obvious to the not so obvious—for example, using a compressed air jet to cool a worker. The air jet might require a flow rate of 20 to 40 SCFM, or 5 to 10 hp of compressed air. The same comfort level can be attained with a $1/4$-hp fan and, unlike the air jet, with very little noise.

Another area of loss is excessive heat. The hotter the air is the less dense it is. As an example, a receiver that is at 185°F and 175 psi will actually have the same amount of air as a receiver at 100°F and 100 psi.

However, significantly more energy, and therefore money, is required to achieve the higher temperature/pressure figure.

Water is also the cause of significant losses in compressed air operations. Two types of losses can be realized from water contamination. First, high water vapor content in compressed air can represent a significant percentage of the air, especially if the air has an excessively high temperature. As the air cools, the water vapor condenses out and the volume of air that the water vapor represented is lost. The second loss, and the one that is better known, is the damage that water can cause to air tools. High water content within the compressed air will accelerate wear on the tool and cause diminished output. In addition to lower power, the life expectancy will be greatly shortened.

Leaks

Air leaks are a constant problem with compressed air systems. The reason that air leaks become such a significant problem is that leaking air hoses don't create a mess like hydraulic or water lines do. Additionally, most air leaks do not represent a safety hazard the way a leaking gas line would. So, unless the leak is large enough to create some sort of problem, most personnel take no notice of it at all. When a workforce harbors a "who cares" attitude toward air leaks, they can grow to the point of being very costly.

Control your leaks! Inspect the system regularly and instruct workers to report any leak they may encounter, regardless of its size. Once it's reported, fix it! Rigorous adherence to a program like this will probably do more to save the company money than anything else in regard to the compression system.

Figure 11-1 shows typical problem areas in a compressed air distribution system. It is possible that threaded pipe fittings will develop leaks over time; however, it is unlikely. When a threaded joint has been installed correctly, it will provide many years of reliable, leak-free service. Threaded joints that are subject to rotation due to expansion and contraction should be inspected on a regular basis.

Much like threaded fittings, flanged fittings offer an excellent leak-free service life. For these joints, 150 lb-raised face flanges with standard plated bolts and a fiber gasket should be used. Do not use gasket sealing compound. After make up, check the joint for leaks with soap bubbles.

Valve stem and gland packings are common points for leaks to develop. All valves in a compression system should be operated and inspected on a regular basis, normally once a year.

Hose terminations, quick disconnects, and universal hose couplers are leak hotspots. These items should be inspected every two months or

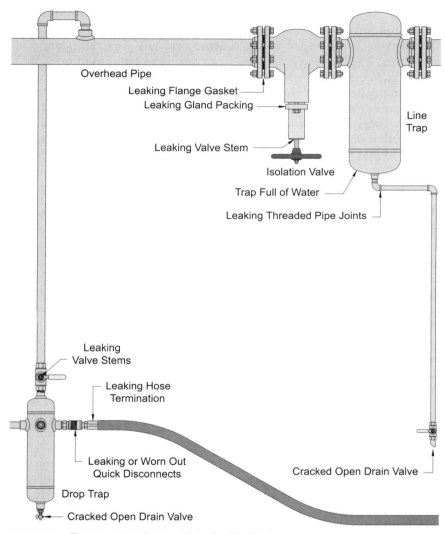

Overhead Pipe

Leaking Flange Gasket

Leaking Gland Packing

Line Trap

Leaking Valve Stem

Isolation Valve

Trap Full of Water

Leaking Threaded Pipe Joints

Leaking Valve Stems

Leaking Hose Termination

Leaking or Worn Out Quick Disconnects

Cracked Open Drain Valve

Drop Trap

Cracked Open Drain Valve

Figure 11-1 Energy saving items within the distribution system.

more. It should be noted that a female quick disconnect generally won't leak when it is not connected, so check it when it is connected as well as disconnected.

Also under the leak category is the practice of cracking open drain valves to continuously drain water. This is a very expensive way of draining water. As an example, if a company has 42 different drain valves cracked open and flowing at 1 SCFM each, then it requires approximately 8.4 hp of compressed air just to feed the drains. This translates to $5856 of energy costs per year! This practice should be aggressively discouraged and a routine draining program should be developed. If it takes too much effort to drain the traps manually, then they may be equipped with automatic electronic drains and strainers. If automatic drains are installed, then the units should be inspected for functionality and their strainers cleaned out at least once a year. In any case, it is very important to drain the traps every day of operation. If a trap is allowed to fill completely it will deliver liquid water to the application. Liquid water can be exceptionally damaging to compressed air components.

Misuse

Be ever on the lookout for misuse of compressed air. When one is found, it should be stopped immediately. It doesn't take very long for an improper use to become normal procedure. Once this happens, it can be very difficult to reverse the trend. One way to restrict misuse is to implement a strict policy requiring authorization from the maintenance group or the plant engineer when connecting any new application to the system.

The Compressed Air Supply

As with the distribution system, the compressor and all associated equipment should be free of leaks. All fittings, joints, valves, and seals should be inspected on a regular basis. The receiver and any traps should be drained daily. This is especially true for the receiver. If the receiver is neglected, then eventually liquid water will be delivered to the distribution system. If the receiver is full enough, then the pump will start to short cycle. Short cycling takes place when the pump turns on and off in rapid intervals and is caused by insufficient receiver volume. During startup, the motor can pull up to ten times its running current. If the start/stop cycle continues to repeat in short intervals, then the motor will overheat and burn up.

Heat

Excessive discharge temperature can contribute to high energy costs. When the air in the receiver is too high, then its density is low. As the air cools, it contracts and the pressure lowers, in effect robbing you of the compressed air that you just paid to generate.

High discharge temperatures can be attributed to two sources, overworking the compressor or a dirty or ineffective aftercooler. The most common problem among smaller reciprocating units is overworking the compressor. A company will purchase a stripped-down compressor that is too small for their application and push it into continuous-run situation. Since the equipment is designed to cool off during off periods, the pump never has a chance to shed the heat of compression and the discharge temperature rises to unacceptable levels. A single-stage compressor may generate temperatures as high as 600°F and a two-stage as high as 400°F. In either case, this is a very ineffective way to generate compressed air. In some cases, the problem may be corrected by converting the compressor to a continuous-run configuration and adding an aftercooler. In more extreme cases, the compressor must be replaced with a larger unit.

Generally, high discharge temperatures in larger screw compressors can be attributed to poor airflow through the aftercooler and, to a lesser extent, the oil cooler. Because of the dirty environments that compressors usually operate in, aftercoolers must be cleaned on a regular basis. It is not unusual to find an aftercooler that has been neglected to the point that there is no airflow whatsoever. If the heat of compression is not removed by the aftercooler, then it goes directly into the receiver. Aftercoolers and oil coolers should be cleaned on a monthly basis, and more often if necessary. It is also important that aftercoolers be located in a well-ventilated area, ideally outside the compressor room. If located inside the compressor room, then the room itself should be well ventilated, preferably with a forced air system.

The Motor

It is not at all unusual to find air compressors over 25 years old. I have seen compressors that were manufactured before World War II supporting significant manufacturing plants every single day. If these old compressors have received proper maintenance, they will most likely provide many more years of service life. The most significant difference between these old compressors and the new ones is the efficiency of the electric motors. Over the past few decades, the efficiency of commercial electric motors has improved dramatically. If an old compressor, which is in good serviceable condition, is equipped with a new "federal" efficiency motor it will produce compressed air at a cost per SCFM that is comparable to a modern unit. In this way a company can lower its cost to produce compressed air without enduring the purchase price of a new compressor.

Screw compressors that are placed into an application with large flow variation can benefit greatly with a variable-frequency motor drive. During low flow periods the motor is slowed down so that the compressor's energy consumption matches the air requirement. In many instances,

equipping a compressor with a motor drive of this nature can significantly reduce the operating costs.

Drive Belts

Many compressors use V-belt drives. It is imperative that the belts be
properly adjusted at all times. Belts that are too tight or too loose will
fail prematurely. Belts that are slipping will force the compressor to operate at a lower output and the system will lose efficiency. The drive belt(s)
should be regularly inspected.

Master Regulator

Most compressed air components are designed to operate at 90 psi.
Therefore, supplying the distribution system with pressures higher
than 90 psi is a waste. Using a master regulator on the output of the
receiver will allow a better utilization of the air charge, and consequently lower cost per SCFM. This is especially true on two-stage units.
It should also be noted that a two-stage compressor can typically deliver
175 psi air. This is a high enough pressure to severely damage most air
tools.

Water

Water contamination must be managed properly to avoid a number of
costly pitfalls. The first is that water vapor can represent a significant
percentage of the volume of the compressed air. This is especially true
when the discharge temperature is unusually high. As the water vapor
condenses, its volume is lost and the pressure of the air charge is
reduced. Consequently the efficiency of the system is reduced. In addition to the loss of volume, water causes other problems in a compression system. The effect of water damage on compressed air components
is discussed in the following section.

Figure 11-2 shows a small reciprocating compressed air supply system
and all the items that should be addressed and/or maintained to lower
energy costs. Pay particular attention to items and conditions that effect
heat. Most of the items shown are duplicated on systems that use screw
compressors.

Water Damage

Water damage is one of the most serious problems facing compressed air
applications. Water within the system will emulsify with lubricating oils
and gum up the internals of delicate components. The gum then attracts

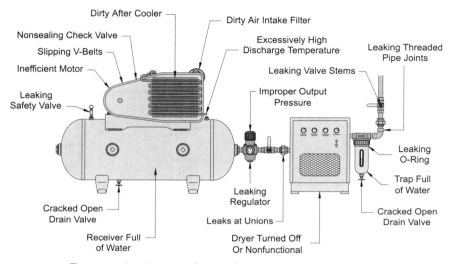

Figure 11-2   Energy savings items at the supply.

and holds particulate contaminants and eventually starts to damage the internals of the system. This is an insidious problem because on the outside, everything looks OK. It's not until the component fails completely that the damage is discovered and by then, it's too late. The only alternative is to rebuild or replace the component. Air motors are especially susceptible to water contamination. The vanes of a motor operate at extremely high speed and rely heavily on proper lubrication. When coupled with the violent nature of the air motor, water is very effective at washing out all the lubricant that is protecting the mechanism. To make matters worse, in the areas where the vanes come in contact with the cylinder, localized heat can build up and vaporize the water, leaving the vane-to-cylinder junction completely dry. At this point, scoring of the cylinder and deterioration of the vanes will occur.

To better illustrate this point, Fig. 11-3 is a graph that shows how the performance of an air tool will deteriorate when it operates on water-contaminated air versus dry lubricated air. The horizontal line represents the power required for the application. The upper and lower lines represent two identical pneumatic angle grinders placed into the same service at the same time. The short line shows an electric grinder placed into the same service as the pneumatic grinders.

The grinder that used dry, lubricated air provided consistent power well into the third year of operation. The performance of the grinder using the wet air started to deteriorate in the first month of operation. By month seven, the power had dropped to match the requirement of

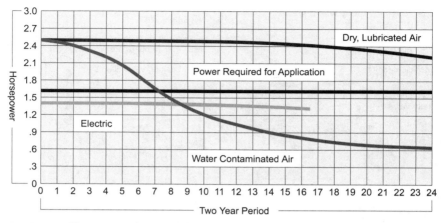

Figure 11-3 Air motor performance deterioration due to water contamination.

the application. By month nine, it had dropped below the performance of the electric grinder. By month thirteen, the power output really wasn't strong enough to be effective in any way. By month nineteen, the operator could stall the grinder with very little effort. The electric grinder provided consistent power until it failed completely in month sixteen. The premature failure of the electric grinder is due to the fact that it really didn't produce enough power for the application. Therefore, the unit was pushed to its limits in an effort to keep up with production. This meant that the electric grinder was operating in an overheated condition most of the time.

The catch is, that the grinder operating on wet air was able to meet the requirements of the application for nearly 9 months and remained fairly effective for another four months. After thirteen months of operation, most companies wouldn't really consider the air tool to be new and therefore, are unaware of the premature failure. In short, the tool drops off their radar screen.

The problems that arise out of this condition are obvious. The power output of the tool drops, production associated with that tool falls off, and rebuild and replacement costs are considerably higher. However, most accountants are unable, or unwilling, to see the value in adding a dryer to the system. This is because they budget an amount for maintenance every month and don't really know what they are paying for. They budget an amount for tool replacement every month, and assume that's just the cost of doing business. Production levels are consistent and there doesn't seem to be a problem there. However, the head of maintenance is standing in front of the accountant asking for $16,000.00 to buy a dryer for no apparent reason. He claims that it will pay for itself, but is unable to outline how. Invariably the accountant will say, "We've never *needed* a dryer before. Why do we need one now?" The response is, "We've never *had* a dryer before. But we've always needed one."

As it turns out, it really is the accountant's responsibility and not the maintenance manager's responsibility to justify the expenditure. The maintenance manager is busy keeping the plant operational and really isn't trained to produce financial reports justifying capital expenditures. However, he is in touch with the plant's requirements and when he recommends a capital expenditure in an effort to save money, the accountant should investigate the request. Unfortunately, most accountants can only see the checks that they're writing and immediately dismiss the request. And a condition that may be costing the company tens of thousands of dollars a year lives on.

Air Versus Electricity

When comparing the power produced from compressed air components with electric components, it becomes clear that compressed air is not a particularly efficient method to transmit energy. In general, a pneumatic component is only 10 percent to 15 percent as efficient as its electric counterpart. This is the reason that very few fixed air motors are encountered in industry. Pneumatic components compensate for their low efficiency with other attributes, principally their power-to-weight ratio. As an example, the electric grinder in Fig. 11-3 is 1.4 hp and weighs 15.5 lb, the pneumatic grinder is 2.5 hp and weighs 8 lb. The electric grinder has a power-to-weight ratio of .09 hp/lb while the pneumatic grinder is .31 hp/lb—a considerable difference.

The overall weight of a pneumatic tool can have a profound effect on the fatigue level that workers experience. As an employee's fatigue increases, his production decreases. Figure 11-4 shows the difference in production between an employee using a pneumatic grinder versus an employee using an electric unit. Notice that in the morning, the production level of the electric unit is 12 percent lower than the pneumatic unit, simply, because it has less power. As the workday progresses, the weight and lower power takes its toll on the worker. By the end of the day, the worker's productivity is 48 percent less than the worker who is using the pneumatic unit. This clearly illustrates that, on paper, the electric unit may seem more efficient; however in the real world it's not.

In addition to high power-to-weight ratios, pneumatic tools have a number of other significant benefits over their electric counterparts. One of the most noteworthy, is that they are inherently safe. Pneumatic tools have no danger of electrical shock, they do not represent a significant fire hazard, can be safely used in wet environments and become cooler during operation, not hotter.

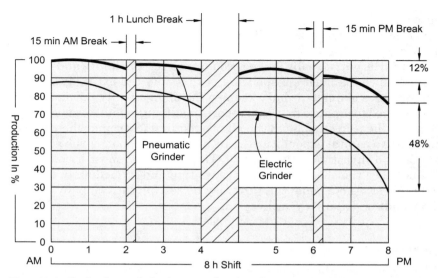

Figure 11-4   Production variation between electric and pneumatic grinders due to fatigue.

Despite this, some shops feel that it is better to avoid compressed air altogether and use more traditional methods to accomplish the tasks at hand. Lets take for instance a furniture shop that operates a numerically controlled router. Figure 11-5 shows the production difference between using a blow-off gun and using a brush to remove the chips and sawdust after each part is complete. The run time on the part is 12 minutes, shown by the solid lines. The time to clean off the finished part is 1 minute with a blow-off gun, whereas it takes 3 minutes with a brush. Production with the brush is 16 percent lower than when a blow-off gun is utilized. Adding a compressed air system may seem like a big expense, but it's a small price to pay to increase production by 16 percent.

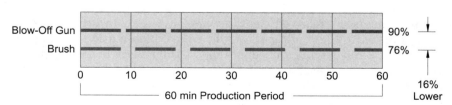

Figure 11-5   Percentage of actual production time during period.

Questions*

1. What is the biggest loss associated with compressed air?
   (A) Misuse   (B) Heat   (C) Leaks   (D) Water

2. What is the second largest loss associated with compressed air?
   (A) Misuse   (B) Heat   (C) Leaks   (D) Water

3. High water content in compressed air can cause what?
   (A) Heat   (B) Diminished output   (C) Spots   (D) Damage

4. The system should be inspected regularly for what?
   (A) Sparks   (B) Paint quality   (C) Leaks   (D) Oil

5. Using a compressed air jet as a fan is considered what?
   (A) A good idea   (B) Misuse   (C) Clever   (D) Efficient

6. What is short cycling?
   (A) When the motor turns on and off in rapid intervals.
   (B) A quick use of compressed air.
   (C) Turning on the compressor for a few minutes.
   (D) Turning off the compressor for a few minutes.

7. Hotter air has lower——?
   (A) Heat   (B) Water   (C) Density   (D) Odor

8. What can hotter air retain more of?
   (A) Liquid water   (B) Water vapor   (C) Oil   (D) Particulates

9. Older compressors should be updated with what kind of motor?
   (A) TEFC   (B) Federal efficiency   (C) Drip proof   (D) Open frame

---

*Circle all that apply.

12

System Models

Most companies operating within a specific industry will have similar compressed air requirements. If your company is a medium-sized manufacturing facility, then it will have compressed air requirements similar to other medium-sized manufacturing companies. This chapter reviews a sampling of typical air compression systems that might be found operating in various industrial sectors.

A contractor's compression system is completely inappropriate for a semi-conductor facility and an instrument air system is inappropriate for a dry cleaner. The question is, where do we start? By reviewing the following models, you can get a fairly good idea of how you might want to design a system. If your particular industry isn't represented here, then review the system that may be closest to your application. As an example, a hospital may use a high purity system, as outlined in the section "High Purity." A machine shop may use a system as outlined in the section "Small Manufacturing Plants." Once you have reviewed this chapter, refer back to information presented earlier in the book to help you configure a system specifically for your application.

The system models that are presented here are intended as a general guideline and not a specific recommendation. It is important to understand that any compression system should be designed specifically for the application at hand.

Contractor Compressors

The primary requirement for a contractor's compression system is portability. The system should be lightweight, easy to operate, and versatile. Major components, which are generally the compressor, dryer(s), and distribution hoses, must be of durable construction and easy to repair.

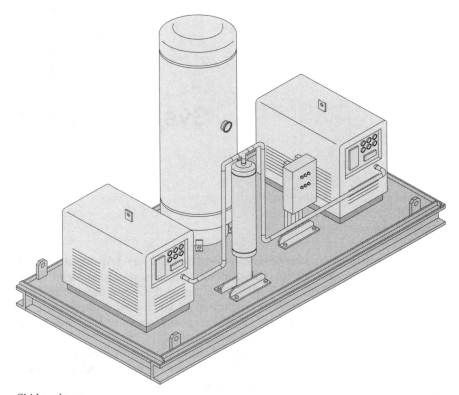

Skid package.

Because these systems are typically set up in the morning and broken down in the evening, all the components should be interconnected with quick disconnects and they should be sized so that they can be easily stored.

Figure 12-1 shows a typical contractor compressor with an ice bath dryer. The compressor should operate on 120 VAC so that power connections are not a problem. The interconnect hose between the compressor should be $^{1}/_{2}$ in. and equipped with $^{1}/_{2}$ in. quick disconnects. Note that the compressor has an output regulator as an integral part of the control panel.

Many contractors operate crews, which rely heavily on pneumatic tools. These crews may be operating at different areas on the job site; therefore, a temporary distribution system must be laid out. Generally, the compressor will be set up near the electrical source and hoses are routed to the various work locations. The primary feed hoses should be large enough to prevent significant pressure drop over the total length.

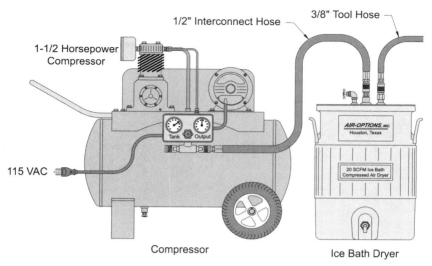

1/2" Interconnect Hose

3/8" Tool Hose

1-1/2 Horsepower Compressor

115 VAC

Tank   Output

AIR-OPTIONS, INC.
Houston, Texas

20 SCFM Ice Bath
Compressed Air Dryer

Compressor

Ice Bath Dryer

Figure 12-1   Contractor compressor.

The primary feed hoses to which the crew can connect can be terminated with a tri-manifold.

Figure 12-2 shows an arrangement that may be deployed at a new home construction site. This particular system has a two-zone distribution system. One zone is for the first floor and the other is for the second floor. Note that there are two ice bath dryers used, one for each zone. This arrangement assures that both zones receive dry air for the entire shift.

Automotive Repair

One of the common uses of compressed air is in the automotive repair industry. It is nearly impossible to find an automotive repair business that doesn't rely heavily on compressed air in their shops. Even weekend mechanics find the use of compressed air critical. The automotive repair industry can be divided into two basic groups. The mechanical group conducts general maintenance on the chassis and driveline. This group includes service stations, brake shops, transmission repair centers, muffler shops, tire dealers, and front-end shops. The second group is the paint and body industry. However, it is not uncommon to find a paint and body shop as a part of a general service shop.

In either case, the air compressor setup is fundamentally the same. Because most automotive repair shops are generally small operations, packaged reciprocating compressors are ideal for these applications.

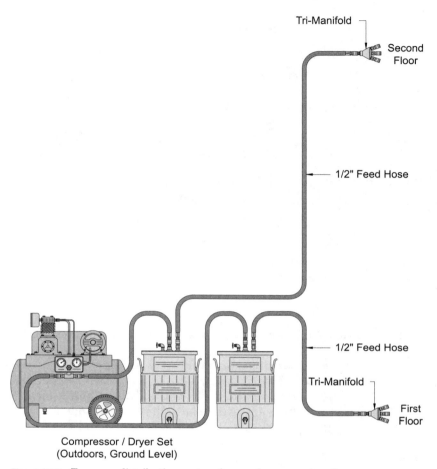

Figure 12-2 Two-zone distribution system for new home construction.

Figure 12-3 shows a two-stage reciprocating compressor with a tank-mounted dryer. The output of the tank is equipped with a master regulator, set at 90 psi, and a line trap. The receiver is equipped with an automatic electronic drain.

Within the mechanical repair shop, air tools are generally handled in the same manner as electric tools. The pneumatic tools are plugged into quick disconnects as they are required. If all the quick disconnects are used, a mechanic will simply unplug the tool he is not using and plug in the one he needs.

In addition to the mechanic's air tool requirement, most auto repair shops have air-over-hydraulic lifts in each bay. These lifts operate in

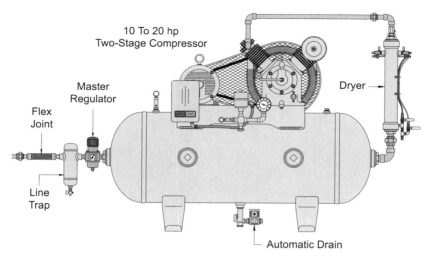

Figure 12-3 Two-stage compressor with dryer.

much the same way as air/hydraulic amplifiers operate. The control for the lift is generally located at a compressed air station on an adjacent wall. Figure 12-4 shows a typical pneumatic workstation with a tri-output trap and hydraulic lift control valve. The valve is simply a three-way valve that is center loaded to off. When the handle is pushed to the up position, air is delivered to the lift. When the handle is pushed to the down position, the air charge in the lift is vented to the muffler. In addition to providing lift control, the workstation provides two quick disconnect ports.

Applying a painted finish to a car body requires a higher level of air quality than other pneumatic applications within the automotive repair shop. If there is even a slight water or oil contamination, the painted surface will be fogged and require substantial buffing to produce the shine that the customer will demand. However, buffing will not correct all the problems that water will cause. The paint will also have contaminants imbedded into the paint itself and the luster will be greatly diminished. The finish will have a chalky or milky appearance that cannot be buffed out. To avoid this problem, the paint booth workstation should be equipped with a canister-type desiccant dryer, a 1-μm particulate filter and an output regulator. Figure 12-5 shows a typical paint booth workstation. The station differs little from the pneumatic workstation shown in Fig. 12-4, with the exception that the lift valve is replaced with the dryer/filter/regulator set.

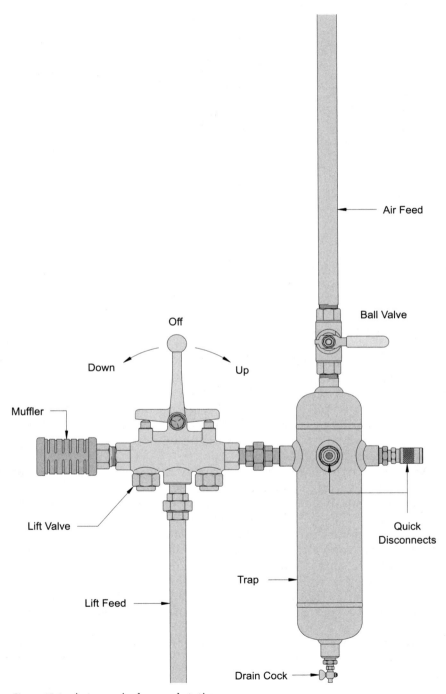

Figure 12-4 Auto repair shop workstation.

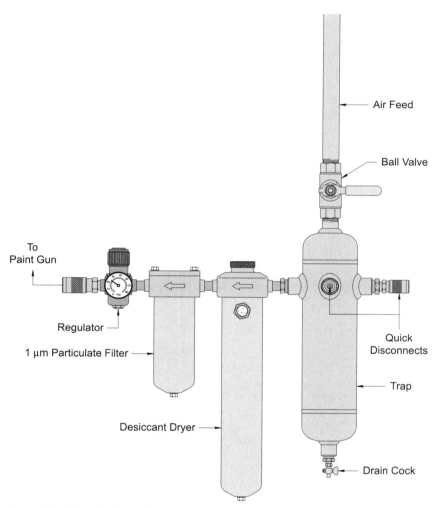

Air Feed

Ball Valve

To
Paint Gun

Regulator

1 μm Particulate Filter

Quick
Disconnects

Trap

Desiccant Dryer

Drain Cock

Figure 12-5   Paint booth workstation.

The distribution system for most automotive repair facilities is a
simple spine arrangement. Figure 12-6 shows a typical example of a four
bay repair shop with two lifts, a front-end bay and a paint booth. The
compressor is generally housed in a room at the back of the shop. All
piping should be located overhead and each drop should be equipped
with a trap. Lifts require an underground connection. This connection
should be limited to the lift feed only. The air supply should be routed
down from the overhead plumbing. Generally speaking, a couple of
quick disconnects should be placed in the front of the shop so that quick
repairs and tire filling can be carried out in the driveway.

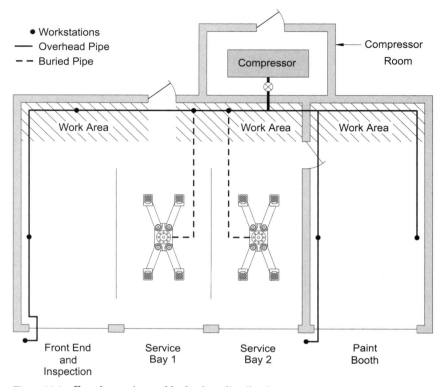

- Workstations
— Overhead Pipe
- - Buried Pipe

Compressor Room

Compressor

Work Area   Work Area   Work Area

Front End and Inspection   Service Bay 1   Service Bay 2   Paint Booth

Figure 12-6 Four bay paint and body shop distribution system.

Small Manufacturing Plants

Small manufacturing applications represent the second most common compressed air installation. This class of system is applicable to a wide range of industries. Any company that has a pneumatic-equipment-intensive operation with 40 to 100 employees will most likely use a system like this. The compression system will typically have two different flow requirements, the high-flow period being normal operating hours and low-flow periods being off-hours and weekends. Many companies operate a second and third shift at a reduced workforce and the air demands during these shifts are much lower. This reduced-demand situation also holds true during weekend operations.

Figure 12-7 shows a typical compression system for a small manufacturing plant. The primary compressor is used during normal operating periods and the secondary, or backup compressor is used during reduced-demand periods. This type of system generally grows as the company grows. The company starts with a small packaged unit and

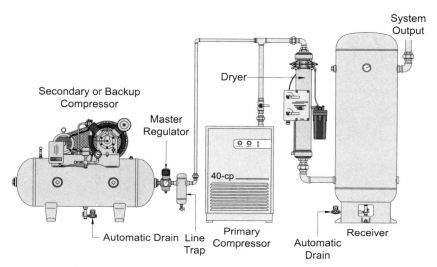

Figure 12-7   Small manufacturing plant compressed air supply.

eventually outgrows its capabilities. To meet the higher demand, a larger compressor is purchased and the smaller unit is placed into secondary duty. In this case, both compressors feed a common dryer that is matched to the larger of the two compressors.

Pneumatic workstations for small manufacturing systems generally have one of three different requirements. The first is the basic quick disconnect; second, dry air applications; third, lubricated applications. Figure 12-8 shows a typical pneumatic workstation, as may be found on any plant floor in the world. Feed lines are hard plumbed directly to the trap or they may be connected through a lubricator, which, in turn, is mounted directly to the trap. It is a good idea to provide quick disconnects at many locations throughout the plant. These disconnects provide easy access for service personnel and allow temporary connections to the system for new or modified equipment.

Pneumatic cylinders and their controls are exceptionally common in automated plant equipment. Equipment that is manufactured and shipped to its final destination will generally have all the pneumatic controls as an integral part of the equipment. The pneumatic controls for equipment that is designed and constructed in-house will generally be mounted remotely from the machine that they serve. This is because remotely mounted controls are very easy to install and service.

Figure 12-9 shows a typical pneumatic workstation carrying a dual-cylinder control arrangement. The regulator provides force control and the four-way valves provide direction control. The valves are wired into a control cabinet, which is connected to the central control system.

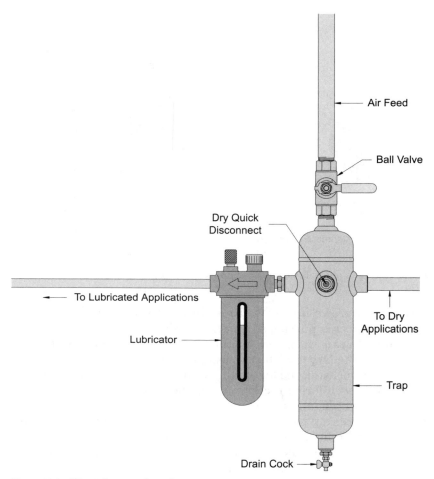

Figure 12-8 Plant floor workstation.

Small manufacturing facilities will usually use a spine distribution system. The spine should be placed at the centerline of the air requirements, usually down the center of the shop. The primary feed should be connected as close to the center point of the spine as possible. The size of the pipe may be progressively smaller as the spine progresses towards the final applications. Isolation valves are placed into the spine so that maintenance can be carried out without disturbing the balance of the system. Workstations are connected to the spine through the application feeds. The compressor is generally located outside the plant in a dedicated compressor room.

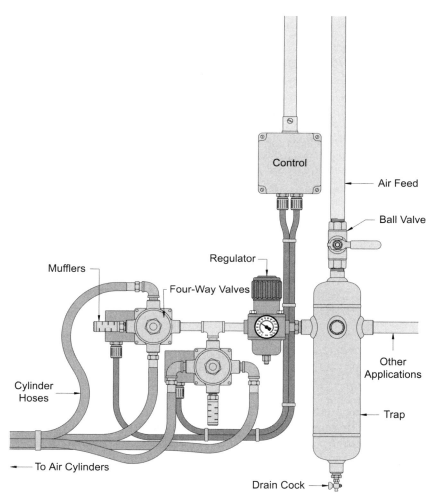

Figure 12-9 Cylinder control station.

Figure 12-10 shows a typical spine distribution system. Note the isolation valves and the graduated pipe sizes used in the spine.

Dry-Cleaning Operations

Pneumatic controls are used almost exclusively on dry cleaning equipment. Any dry cleaning operation will have a compression system feeding control air to the facility. Because the compressed air is used primarily for control, even relatively large operations will have rather

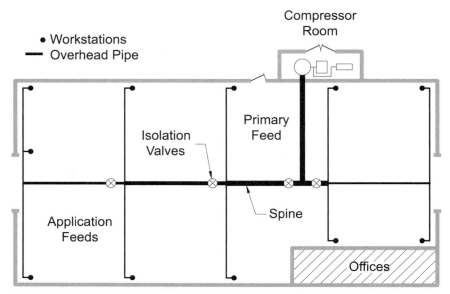

Figure 12-10   Spine distribution system for small manufacturing plant.

modest compressors. If system leaks are carefully controlled, then a 5 or 10-hp compressor can support a fairly large operation.

Figure 12-11 shows a typical control air compressor, which is applicable for dry-cleaning operations. Single-stage compressors are generally selected because they are less expensive, less complex, and easier to maintain. The control pressure is generally in the 35- to 80-psi range and a single-stage compressor can easily deliver this. The compressor should be set up with a tank-mounted dryer, master regulator, line trap, and automatic drain, as shown.

The pneumatic workstations used in dry-cleaning operations generally need to deliver either lubricated air or dry air. Usually system pressure is set at the master regulator (80 psi) and secondary regulators are set up for each piece of equipment.

Figure 12-12 shows a typical pneumatic workstation for supporting dry-cleaning equipment. A trap is set up with a regulator and lubricator set, for equipment that requires lubrication, while only a regulator is used for nonlubricated applications.

A compressed air requirement needed by many dry cleaners is an airbrush. These devices are used to touch up blemishes on leather, vinyl, and other specialty fabrics. A dry cleaner that is skilled in matching and applying colors with this method can effect remarkable repairs to garments that would otherwise be discarded. Generally, a dry cleaner will have a special bench set up for airbrushing. The pneumatic workstation

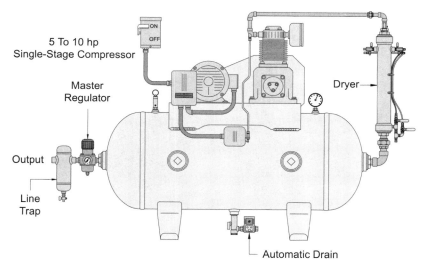

5 To 10 hp
Single-Stage Compressor

Master
Regulator

Dryer

Output

Line
Trap

Automatic Drain

Figure 12-11   Single-stage compressor with dryer.

should be equipped with a desiccant dryer, 5-μm filter and a regulator. A dust-off gun is also handy. The regulator is set to a pressure that is appropriate for the airbrush.

Figure 12-13 shows a typical airbrush workstation.

Air distribution for a small dry cleaning facility is not a difficult proposition. A "C" loop is generally installed, which corresponds with the location of the pneumatic equipment. Each workstation should be equipped with a trap and located in a position that allows easy access. The compressor is generally located outside, in a dedicated compressor room.

Figure 12-14 shows a typical compressed air distribution system for a small dry cleaning operation.

Heating, Ventilation and Air Conditioning (HVAC)

Pneumatic controls are used extensively in the Heating, Ventilation and Air Conditioning (HVAC) industry. Most large commercial buildings control their heating and cooling equipment with some sort of pneumatic controls. Many of us have encountered pneumatic thermostats in a commercial building and thought, "what the heck is that and how do I adjust it?" Pneumatic controls offer a simple and reliable way of controlling the intricacies of very large, centralized heating and cooling systems. The heating and cooling system generally has a temperature-controlled water circulation system, which is interconnected to various air

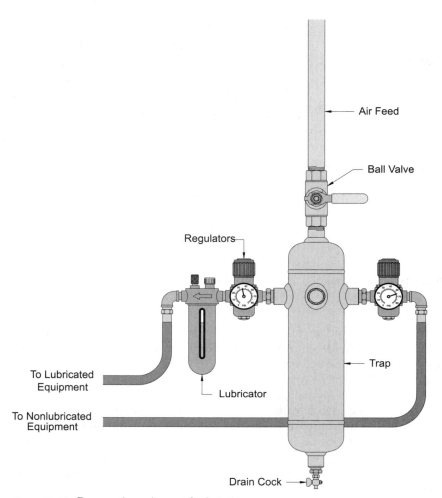

Figure 12-12 Pneumatic equipment feed station.

handlers throughout the building. The functionality of these air handlers and their flow characteristics are generally controlled by pneumatics. In the simplest form, a pneumatic thermostat will control a diaphragm cylinder, which, in turn, controls a damper valve within the ducting.

The compressed air source is particularly important in these systems. If the compressor fails, then the heating and cooling system shuts down for the entire building. For this reason, duplex compressors are almost always specified for these applications. Figure 12-15 shows a typical duplex compressor setup for HVAC applications. The compressor should

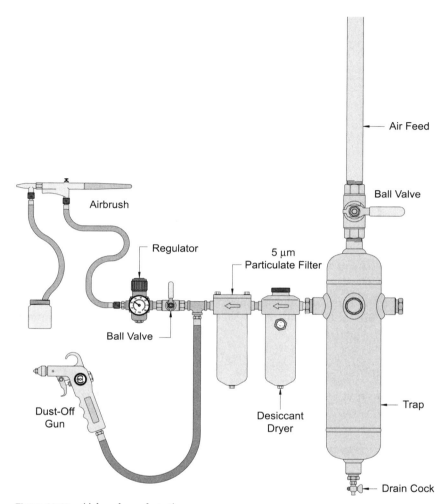

Figure 12-13 Airbrush workstation.

be a dual pump, packaged unit equipped with a tank-mounted dryer, master regulator, line trap, and automatic drain.

HVAC control pressure ports typically consist of a three-way venting valve, control pressure regulator, and a distribution manifold. The control pressure is set between 20 and 30 psi and the controls are generally connected with plastic hoses.

Figure 12-16 shows a typical control pressure port. The main line pressure is usually set by the master regulator at 60 psi. Dampers are usually spring-loaded in the closed position. When air pressure is applied, the diaphragm cylinder opens the damper. The thermostat vents air in

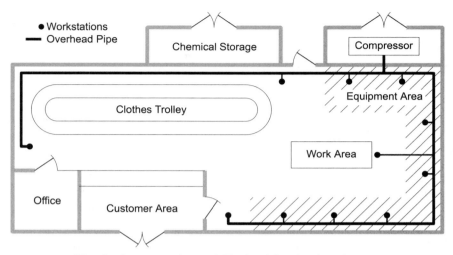

Figure 12-14 Distribution system for a neighborhood dry cleaning plant.

reference to its temperature setting and reduces the delivered pressure in order to control the position of the damper, and consequently the air flow through the duct.

Dental Systems

Anyone who has been to the dentist's office knows that compressed air plays an important role in dentistry. Modern dental drills are

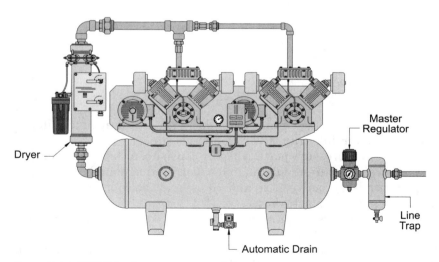

Figure 12-15 HVAC duplex compressor and dryer.

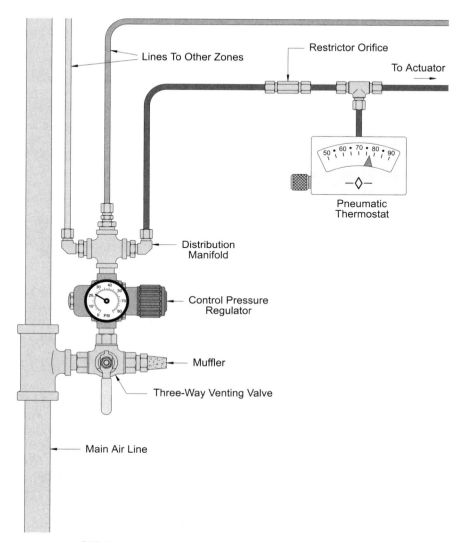

Figure 12-16 HVAC control pressure port.

indispensable tools of the profession. What most people never consider is the source of that air. A properly configured and maintained dental compression system will provide reliable, safe compressed air for many years.

Figure 12-17 shows a typical dental compressed air system. The compressor is a packaged rotary vane unit that is equipped with a master regulator and micro-grid coalescing filter. A flexible line is used to connect the

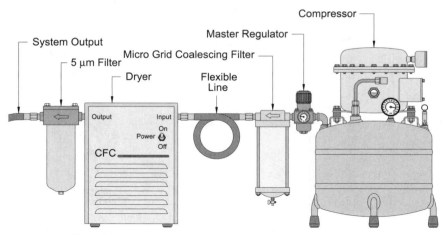

Figure 12-17 Dental compression system.

output of the filter to a CFC-based dryer with a 5-μm particulate filter on the output. The system output is plumbed directly into the distribution system. It should be noted that these compressors should use food service oil because the compressed air is discharged into the patient's mouth.

Home Shop

For many hobbyists, a compressor in the back of the garage is a dream come true. These systems are fairly simple and inexpensive to purchase. A good, solid home shop compressor is typically a single-stage unit in the $^3/_4$- to 2-hp range. Compressors that operate on 120 VAC add a level of convenience by allowing the unit to be simply plugged in. The system should be set up to turn on at 110 to 115 psi and turn off at 120 to 125 psi. The output of the tank should be equipped with a master regulator that is set to deliver an output pressure of no higher than 90 psi. Following the regulator should be a coalescing filter with suitable quick disconnects. An item of great convenience is a tri-manifold. This item will provide you ready access to three quick disconnects.

Figure 12-18 shows a typical home shop compressor and accessories. When properly maintained, a unit like this will provide a lifetime of service.

Medium to Large Manufacturing Facilities

Compression systems for medium to large manufacturing facilities are not particularly complicated. These compressors are simply bigger versions of smaller systems. Generally, a large system will consist of several

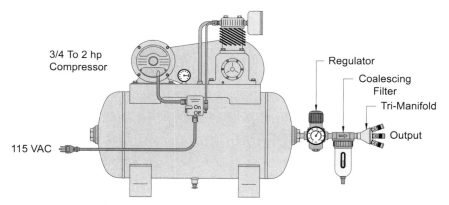

Figure 12-18 Home shop compression system.

compression sites that feed a common loop distribution system. Each compressor site will be configured as shown in Fig. 12-19. There will be two compressors set up in a redundant installation. The compressors will feed a single refrigerated air dryer which, in turn, feeds the receiver. Any substantial manufacturing facility should have at least two compressor sites. This provides a high degree of redundancy and guarantees that the plant will not be forced to endure either scheduled or unscheduled interruptions.

Pneumatic workstations should be standardized within any large facility. This allows the plant engineers to easily access any modifications or expansions planned for the system. In addition, this allows maintenance to stock a series of standard parts so that any problems or modifications can be carried out in short order.

Figure 12-20 shows a typical workstation equipped with a tri-output trap. The input of the trap is equipped with a ball valve, which allows the applications or secondary distribution system to be easily isolated for maintenance. It's also a good idea to equip these workstations with a quick disconnect for service or temporary applications.

Isolation valves and line traps can be very important in large loop distribution systems. Because it is difficult to maintain a reasonable drain back angle over long distances, line traps are placed throughout the system to trap any water buildup. When installing a line trap in an overhead distribution loop, plumb the drain line and valve down to floor level.

Isolation valves are extremely important in a large system. Isolation valves allow sections of the system to be turned off and vented without disturbing other operations. When designing or reviewing a distribution system, carefully consider locations for isolation valves.

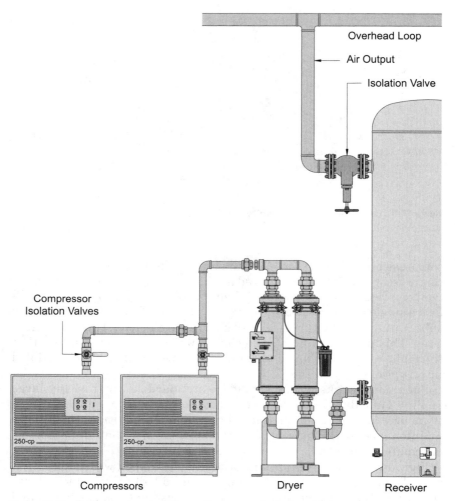

Figure 12-19   Redundant system for medium manufacturing plant.

Figure 12-21 shows a typical isolation valve/line trap installation in an overhead distribution loop. Also notice the configuration of the drop connection. Taking an air-drop off of the top of the loop provides another level of water trap.

Distribution systems for medium to large manufacturing plants are generally laid out in a loop configuration. For large facilities, a loop system has several advantages. The first being that the system naturally equalizes pressure at any given location. An application drawing an unusually large amount of air would have a tendency to deprive

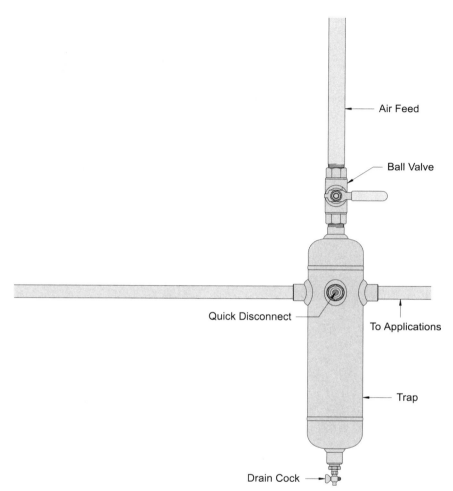

Air Feed

Ball Valve

Quick Disconnect

To Applications

Trap

Drain Cock

Figure 12-20 Plant floor workstation.

other applications that are further away from the compressor in a spine system. The loop system, however, will make up for these deficiencies by feeding air from the opposite side of the application. In a sense, it provides two sources of air for every application connected to the loop. Another attribute of a loop system is that the center of the building remains free of overhead obstructions. This allows the unimpeded operation of overhead cranes.

Figure 12-22 shows a schematic of a loop-type distribution system. The balance pipes are added to further equalize the pressure in the primary loop. Take note of the number and location of isolation valves and line

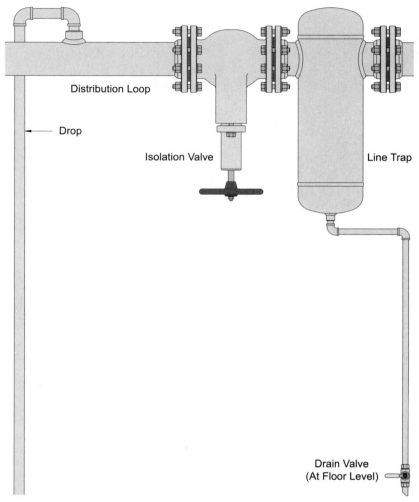

Distribution Loop

Drop

Isolation Valve

Line Trap

Drain Valve
(At Floor Level)

Figure 12-21   Overhead isolation valve and line plant.

traps. Line traps are applied anywhere a suitable drain back angle cannot be maintained.

Instrument Air

Chemical plants generally have two air quality specifications—utility air and instrument air. Utility air has the same basic quality as general purpose air in any other application. Instrument air has a much more stringent specification. Many plants arbitrarily adopt a specification of −40°F

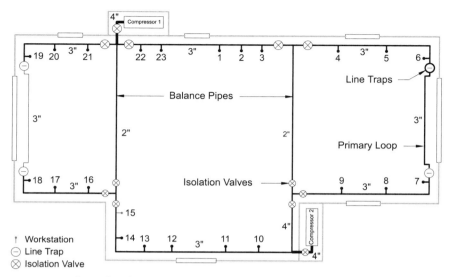

Figure 12-22 Loop distribution system.

dew point and a pressure regulation within ±0.5 psi, however, this specification may vary from plant to plant. Some plants specify their dew point as X degrees below the lowest temperature recorded in the previous Y years. As an example, a plant on the Gulf coast may have an instrument air specification of 20°F below the lowest temperature recorded in the previous 30 years. If the lowest temperature was −8°F, then the instrument air specification will be −28°F. In real life, instrument air systems rarely meet their dew point or pressure regulation specifications. These stringent specifications are rarely necessary and are adopted or assigned by personnel that have no understanding whatsoever of compressed air systems. Most instrument air systems I have inspected are using dryers that are too small for the flow requirement, are not functioning properly or are bypassed altogether. I have had plant engineers proudly give me tours of their instrument air systems, extolling the quality of the air output while standing in front of a desiccant dryer in bypass mode. The actual quality is no better than the utility air output. Another telltale sign is that most instrument air systems have no air quality monitoring equipment. This is primarily because maintaining these rigorous specifications requires constant maintenance.

Figure 12-23 shows a typical instrument air compression system. The system consists of two compressors set up in a redundant configuration. The output of the compressors are fed through a check valve and into a tank-mounted Elliott cycle dryer. The output of the dryer feeds the utility air receiver which, in turn, feeds the utility air output. The utility

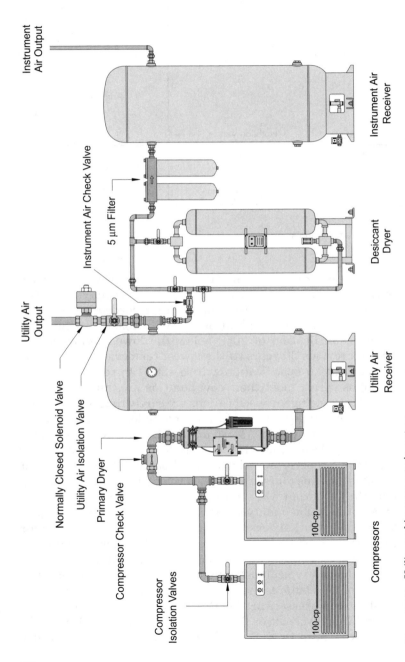

Instrument Air Output

Instrument Air Check Valve

5 μm Filter

Normally Closed Solenoid Valve

Utility Air Isolation Valve

Primary Dryer

Utility Air Output

Compressor Check Valve

Compressor Isolation Valves

Instrument Air Receiver

Desiccant Dryer

Utility Air Receiver

Compressors

100-cp

100-cp

Figure 12-23 Utility and instrument air system.

air output is equipped with a normally closed solenoid valve that will shut off the utility air flow in the event of a power failure. This is intended to preserve as much air as possible in an effort to maintain the instrument air output. Utility air is generally at 38°F to 55°F dew point and is used for all general purpose applications. The output of the utility air receiver is also directed through a check valve and into the desiccant dryer. The output of the desiccant dryer feeds a 5-μm twin tower particulate filter, which is mounted on the receiver. The output of the receiver is plumbed into the plant. The instrument air receiver is generally oversized for the application for the same reason that the solenoid valve is placed on the utility air output: to guard against power failure.

Specifying an instrument air system is no simple matter and a great deal of time and study should be applied before specifying the system components. It is rare to find a compressed air dealer that has sufficient expertise in this regard, so when hiring an outside consultant, ask for and carefully scrutinize his or her credentials. Always remember, the primary duty of a dealer is to sell the products and services that he represents.

Chemical plants typically cover many acres of ground and in some instances hundreds of acres. Some plants are bigger than small towns, and, as such, have the same utility problems that a small town may have. Most plants deal with the plant's requirements by utilizing a utility trunk. The utility trunk will provide ready access to the essential utilities that any chemical plant may require. A typical trunk will have electrical, control, instrument air, utility air, cooling water feed, cooling water return, and waste-water loops. The utility trunk in many plants will also have specialty utilities as well as the standard ones.

Figure 12-24 shows a typical utility trunk for a chemical plant. Note that isolation valves and line traps are carried within the trunk.

Chemical plants are generally laid out in units. Each unit processes a specific product or provides a specific service. Compression systems are usally set up and assigned their own unit number. The distribution system is generally a primary loop with secondary loops in each unit, as shown in Fig. 12-25.

High Purity

High purity compression systems are specified for applications that require extremely clean air, most notably, semiconductor manufacturing. The quality of air that is plumbed into the fabrication laboratories must be very dry and exhibit exceptionally low particulate and hydrocarbon contamination. The dew point requirement specified by most facilities is −80°F, the particulate specification is 0.001 μm and the hydrocarbon specification is less than one part per million. Achieving these specifications requires a carefully designed system. Maintaining

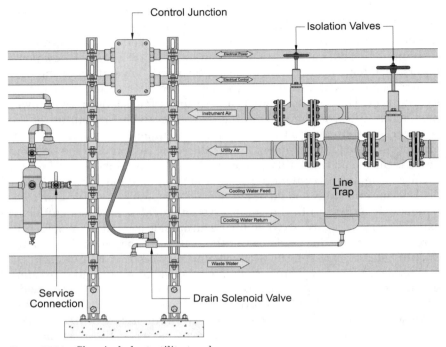

Control Junction

Isolation Valves

Electrical Power

Electrical Control

Instrument Air

Utility Air

Cooling Water Feed

Line Trap

Cooling Water Return

Waste Water

Service Connection

Drain Solenoid Valve

Figure 12-24 Chemical plant utility trunk.

these specifications requires a rigorous maintenance and monitoring program. Unlike instrument air applications, these facilities really do need this level of air quality.

Compressed air in these facilities is used for all manner of control, from operating cylinders to providing flow for air bearings. Additionally, it is commonly used for ultraclean dust-off operations. The air quality is required because of the microscopic scale of the semiconductors being manufactured. When these components are examined under a micro-scope, a 1-μm particle looks like a 20-ft boulder lying in the middle of the street. A 1-μm drop of oil is a greasy flood, which will completely destroy the circuit.

A high purity system starts like any other system, with a pair of com-pressors set up in a redundant installation. Many engineers specify oil-less compressors in these installations, however, there is some debate as to whether an oil-less or a lubricated compressor is better. Naturally, oil-less compressors have significantly lower hydrocarbon contamination on their outputs, however, they achieve this at the expense of other parameters. An oil-less compressor will operate at a much higher speed than its lubricated counterpart. Coupled with the higher speed and the

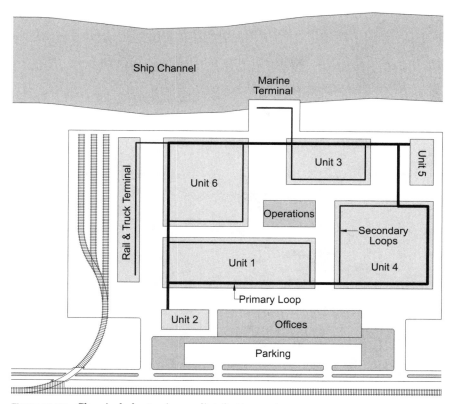

Figure 12-25 Chemical plant primary distribution system.

lack of cooling from lubrication oil, the discharge temperature from an oil-less compressor is much higher and, therefore, it carries a significant higher water load. Additionally, the lack of lubricants coupled with high speeds and close tolerances means that the output will have a particulate contamination, usually metal particles. Lastly, oil-less compressors exhibit a higher purchase price and have a considerably lower life expectancy than their lubricated counterparts. The debate over oil-less versus lubricated is centered around the idea that if you can filter out a higher water and particulate load, then you can filter out hydrocarbon contaminants. Therefore, the considerably higher expense and lower life expectancy of an oil-less compressor is not warranted.

Figure 12-26 shows a high purity compression system that may be found in a medium-sized semiconductor facility. The compressors feed floor-mounted refrigerated air dryer. The output of the refrigerated dryer is fed directly into an oversized twin tower desiccant dryers. The output of the desiccant dryer feeds a 5-μm twin tower particulate filter.

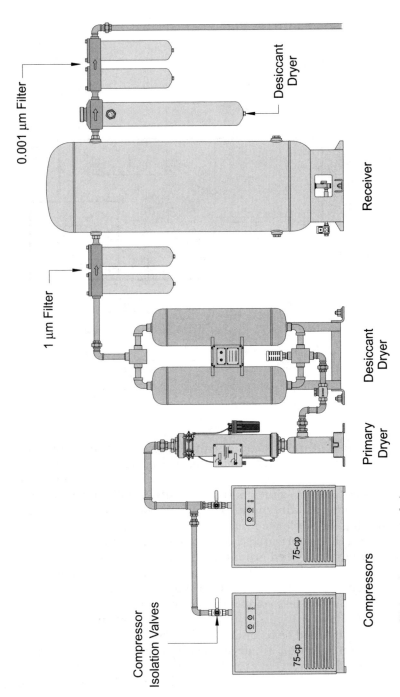

Figure 12-26 High purity compressed air system.

At this point the air introduced into the receiver should be in the $-40°$F to $-60°$F dew point range. The output of the receiver is processed through a secondary canister-type desiccant dryer that is sized to polish the air to $-80°$F. The output of the secondary dryer is processed through a final 0.001-μm filter set.

High purity workstations must be constructed using only high purity components. Additionally, these components should undergo a detailed cleaning before being installed. It makes no sense to install a compression system as described in Fig. 12-26 if you're going to use hardware store components at the application site.

Figure 12-27 shows a typical high purity workstation. They will typically consist of a ball valve, regulator, flow meter, 0.001-μm line filter and output manifold. The 0.001-μm line filter is intended to provide redundancy. For equipment that requires several different delivery pressures, a bank of workstations may be installed.

In order to maintain the degree of air quality that is necessary in high purity applications, the delivered air must be sampled and monitored. Most companies set up an air analysis system that constantly monitors and records the various parameters that are considered critical. In most cases these parameters are dew point, hydrocarbon content, and pressure regulation.

The instrument shown in Fig. 12-28 is set up to monitor the above parameters and provide a continuous paper printout as well as communicating to a central computer system. The unit is constructed within

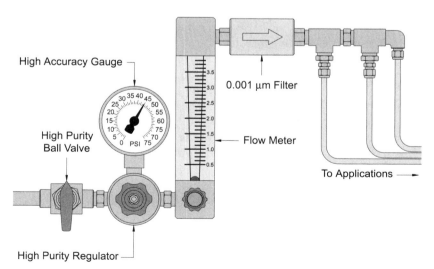

Figure 12-27 High purity workstation.

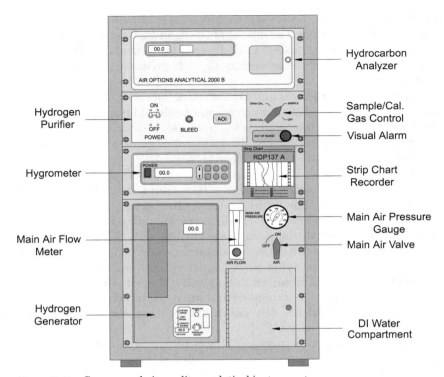

Figure 12-28   Compressed air quality analytical instrument.

a standard 19 in. EIA rack cabinet. The top unit is the hydrocarbon analyzer. The hydrogen purifier, hydrogen generator, deionized (DI) water source and sample/cal. gas controls are required to support the hydrocarbon analyzer. The hygrometer measures dew point and the main air-flow meter is used to adjust the continuous throughput of the instrument. The main pressure gauge reads the system input pressure and is equipped with a 0- to 10-VDC output. The hydrocarbon analyzer and hygrometer are also equipped with a 0- to 10-VDC output. The outputs of all three instruments are connected to the strip chart recorder, which also has a digital output. The digital output is set up to communicate with a central computer system.

Service Trucks

Field service operations exist in many industries and, for the most part, these operations are based on light trucks set up for this specific duty. Many of these service trucks carry a small compression system intended to deliver air for the various pneumatic tools that the service technicians may use. The compressor is generally a gasoline powered, two-stage unit mounted on an undersized receiver. The receiver is designed specifically for this duty and is generally a low profile unit with heavy-duty feet. Another attribute of these receivers is that the tank drain is typically a pick-up tube with an output on the top of the tank.

Figure 12-29 shows a typical service truck compression system. The output of the tank is equipped with a master regulator, coalescing filter, and lubricator. There are two different outputs—a dry one and a lubricated one. The two different outputs should carry different size quick disconnects so that the tools are not misconnected. It should also be noted that, when selecting a service truck compressor, it's a good idea to specify an electric start on the engine. This feature makes the task of starting the compressor considerably easier on cold winter mornings and hot summer afternoons.

One of the most common applications for service trucks is roadside assistance for tractor trailer rigs. Most of us have seen these trucks parked next to a disabled rig as we drive by. The tools required for a truck of this nature are fairly basic. Figure 12-30 shows a selection of tools for this application. The $^3/_4$- and $^1/_2$-in. air wrenches use a larger quick disconnect that matches the $^1/_2$-in. hose and lubricated output on the compressor. The tire inflator and blow-off gun use a smaller quick disconnect which matches the dry air output on the compressor. Another handy part to have is a tri-manifold, which allows the inflator and blow-off gun to be connected at the same time.

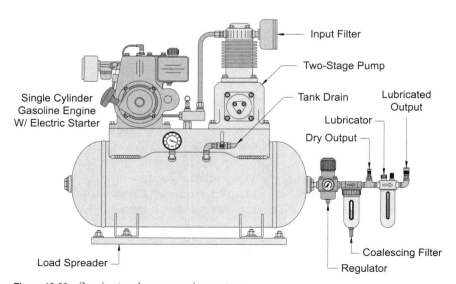

Figure 12-29 Service truck compression system.

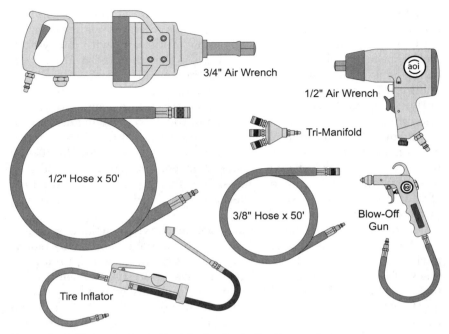

3/4" Air Wrench

1/2" Air Wrench

Tri-Manifold

1/2" Hose x 50'

3/8" Hose x 50'

Blow-Off Gun

Tire Inflator

Figure 12-30   Various tools required for tractor-trailer tire service.

Field and Construction

Field service operations, heavy construction, and road crews oftentimes deploy large mobile compression systems. We have all seen and heard road crews using jackhammers to break up concrete to effect repairs. Compressed air used in field applications is usually delivered from a trailer-mounted construction compressor. These units are very common and are readily available in sizes ranging from 120 to 1900 SCFM. These compressors are typically screw compressors, which are driven with a small industrial diesel engine. The compressors are usually rather spartan and either do not have an aftercooler or have a very small unit which is an integral part of the engine's radiator. Because of this, the output air quality of these compressors is generally rather poor. It will be at a high temperature and carry a heavy oil and water content.

Figure 12-31 shows a typical trailer-mounted construction compressor. These units are usually compact and easily towed behind an ordinary pick-up truck.

Distribution systems to support field operations may be as simple as a single hose connecting the compressor with a jackhammer or it may be an extensive system of hoses fed by multiple compressors. In most cases, field systems use universal hose fittings for all interconnections.

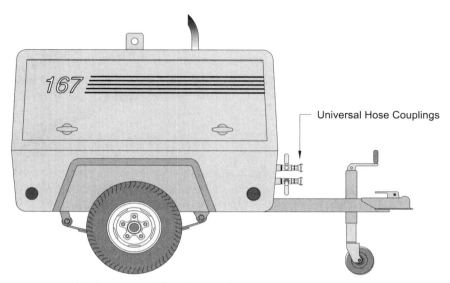

Figure 12-31   Trailer mounted diesel-powered screw compressor.

Figure 12-32 shows a few basic items that any field operation will be able to utilize. Adaptors are very useful to have on hand. They provide a convenient way of connecting multiple hoses of different sizes. Distribution manifolds are a must for larger systems. These units provide workers ready localized access points to the compression system. Universal hose couplings to quick disconnect adaptors provide an interface from the primary distribution system to common pneumatic tools. When setting up a large distribution system, the primary loop should be laid out with 2-in. hoses interconnected with distribution manifolds. The secondary feed lines should be 1-in. hoses.

Most types of dryers are available in a configuration for field operations. The problem is that preparing a dryer for this type of service is very expensive and the dryer rarely stands up to the abuse that it receives.

The only dryers that are generally heavy duty enough to provide longterm service in this arena are Elliott cycle dryers. These dryers are rather compact, are of all welded construction, and have no moving parts. This makes them ideally suited for the rough handling that the equipment invariably receives. Figure 12-33 shows a 2000 SCFM trailermounted Elliott cycle dryer.

Figure 12-34 shows how a field distribution might be laid out. This is an arrangement for sandblasting and repainting a tank farm. The primary loop, which encircles each tank is 2-in. hose. The hosed are interconnected with 2-in. three-way crosses and a distribution manifold

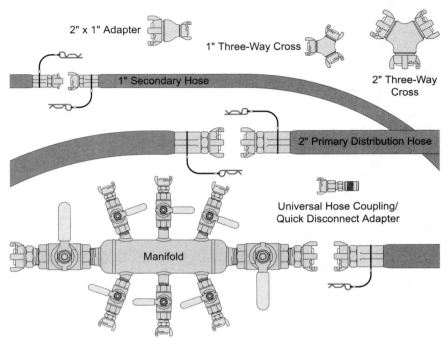

Figure 12-32 Field distribution components.

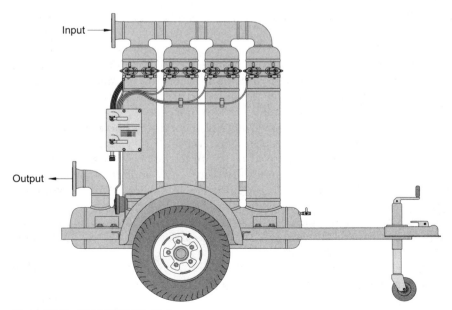

Figure 12-33 2000 SCFM field dryer.

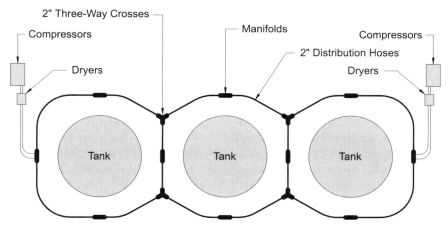

Figure 12-34   Field distribution layout.

is placed on each quadrant of the tanks. The compressors feed the loop through two 1-in. lines, each connected to the end distribution manifolds.

SCUBA and SCBA Charging Systems

Air pressures used in SCUBA and SCBA applications are typically in the 2000- to 4500-psi range. In order to generate these levels, a special high-pressure compressor is required. Additionally, all materials, oils, filters, and equipment must be rated for breathing air.

Figure 12-35 shows a typical four-stage piston compressor, which can produce 5000-psi output. The pumps are usually radial designs with simple

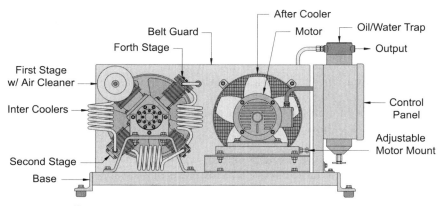

Figure 12-35   Four-stage 5000 psi breathing air compressor.

intercoolers between each stage. The output of the fourth stage feeds a high-efficiency aftercooler. The aftercooler is generally pumped by a motor-mounted fan, as shown. The advanced aftercooler is required because of the amount of heat generated when compressing air to such high levels. The output of the aftercooler is routed through a high efficiency oil/water trap. These types of compressors are normally supplied as base-mounted units and, if required, a receiver is added during installation.

Figure 12-36 shows how the entire supply system would be laid out. The output of the compressor is plumbed through a high-pressure CFC dryer. The output of the dryer is sent through a cartridge-type desiccant dryer, which, in turn, feeds a twin tower particulate filter set. The filter elements should be 1 μm or finer. The output of the filter set is directed to the receiver. It should be noted that all components in the system must have a working pressure of 5000 psi or greater.

The receiver is generally a double-ended, spun, high-pressure cylinder. The cylinder must be equipped with a rupture disk that is rated to fail at 200 psi over the maximum system pressure. Isolation valves are placed at both the input and output of the receiver. The output is also equipped with a master regulator, which is generally set at 3000 psi for SCUBA and 2250 psi for SCBA. All pipe used in the system should be schedule 80, seamless, type 316 stainless steel, or heavier. Pipe fittings should be 3000 psi type 316 stainless steel. Because standard pipe fittings can be difficult to use at these pressures, seamless type 316 stainless steel tubing and compression fittings are often used. If tubing and compression fittings are used, be certain that they are rated for the maximum system pressure.

Figure 12-37 shows a typical filling for SCUBA tanks. Tanks are placed into a water tank and connected to the manifold through high pressure hoses and tank yokes. The water tank is used to help cool the tanks during filling. The manifold should have a pressure gauge, main valve, and a vent. The main valve is intended to isolate the manifold from the supply. The vent is for venting tanks with unknown mixtures. Oxygen cylinders are plumbed into the manifold for mixing nitrox, air with a higher percentage of oxygen.

When filling tanks, it is preferable that a defined procedure be used. In this manner the filled tanks, whether straight air or nitrox, will be consistent. The station should be set up with an empty rack, water tank, and full rack in a left to right progression. If nitrox is mixed, then three full racks should be set up. One for air, one for 25 percent mix, and one for 38 percent mix. The manifold should be placed above and behind the water tank, usually mounted to a wall. Only personnel that are properly trained and certified to fill tanks should conduct filling procedures. If the charge pressure or mixture is in doubt, vent the tank and recharge. Nitrox mixtures should be clearly labeled with the charge

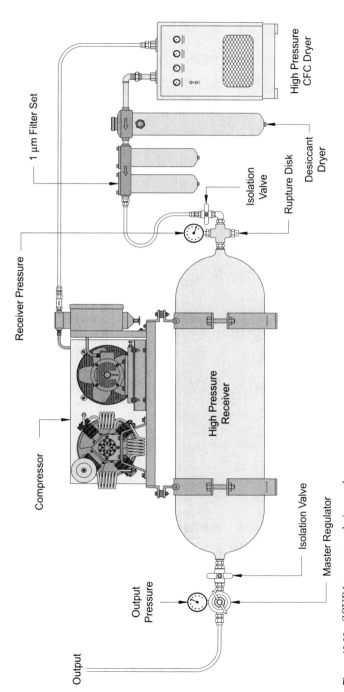

Figure 12-36 SCUBA compressed air supply.

Output

Output
Pressure

Compressor

Receiver Pressure

1 μm Filter Set

High Pressure
CFC Dryer

Desiccant
Dryer

Rupture Disk

Isolation
Valve

High Pressure
Receiver

Isolation Valve

Master Regulator

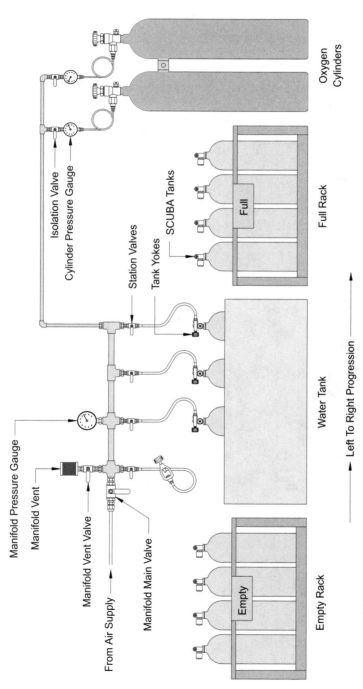

Figure 12-37 SCUBA tank filling station.

Manifold Pressure Gauge

Manifold Vent

Manifold Vent Valve

From Air Supply

Manifold Main Valve

Isolation Valve

Cylinder Pressure Gauge

Station Valves

Tank Yokes

SCUBA Tanks

Oxygen Cylinders

Full

Full Rack

Water Tank

Empty

Empty Rack

Left To Right Progression

318

date, pressure and percentage of oxygen. The O-ring on each tank valve should be inspected before the tank is connected for filling. O-rings that are dirty or show any damage whatsoever should be replaced before filling.

The entire system should be maintained in top condition at all times. Be certain to conduct all of the manufacturer's recommended service on major components. Inspect the piping system and valves on a regular basis. Verify that all of the pressure gauges are accurate and functioning properly. Maintain a good stock of replacement filters and replace them on a scheduled basis. Allow only properly trained personnel to conduct maintenance and inspections. It's also a good idea to keep a log of all inspections and maintenance. And finally, keep the system clean and free of any clutter.

It should be noted that a dive shop with a charging system in a poor state of repair is taking a chance with your safety, not theirs. Ask to see the charging system that your shop uses; if it looks questionable, maybe it's time to find a new shop.

Questions*

1. What type of dryer should contractors use?
   (A) CFC   (B) Ice bath   (C) Desiccant   (D) Traps

2. What type of compressor is best for small automotive shops?
   (A) Screw   (B) Rotary vane   (C) Turbo   (D) Reciprocating

3. What type of distribution system does a small manufacturing plant use?
   (A) Spine   (B) Loop   (C) Hose   (D) Copper

4. Dry cleaners require what two applications?
   (A) Controls   (B) Spray painting   (C) Air wrenches   (D) Airbrush

5. What is the typical control pressure for HVAC systems?
   (A) 175 psi   (B) 100 psi   (C) 20 to 30 psi   (D) 60 psi

6. What voltage do most home compressors operate on?
   (A) 120 VAC   (B) 120 VDC   (C) 240 VAC   (D) 480 VAC

7. What type of isolation valve should be used in large distribution systems?
   (A) Globe   (B) Ball   (C) Gate   (D) Butterfly

8. What arbitrary dew point do plants specify?
   (A) 55°F   (B) 38°F   (C) –80°F   (D) –40°F

9. What power source do most field service compressors use?
   (A) Diesel   (B) Gasoline   (C) Propane   (D) Steam

10. What type of hose connector is used for large field operations?
    (A) Industrial   (B) SNPT   (C) Barbs   (D) Universal

*Circle all that apply.

13

Specifying a Compression System

At this point you should have a fairly good understanding of air compressors, dryers, tools, plumbing, and systems. With this basic knowledge and the information contained in this chapter, you will have the necessary expertise to specify a compression system that is appropriate for the application that is facing you. If you are expanding an existing system, it's a good idea to conduct an air audit first. However, it is recommended that you review all the information in this chapter before conducting an audit. Chapter 14 provides a detailed discussion of air audits.

The very first thing to consider is how the compression system will be used. Think about workstations, plumbing routes, compressor location, compressor noise, drain positions, ease of maintenance, future expansion, and the like. Spend some quality time on this topic. Make notes on everything that you consider important. And above all, don't take it lightly! Doing a cut-rate job will come back to haunt you or your replacement one day. Put in plenty of drops, even in places where you don't think they are necessary. Place an air port at the back dock so that employees and delivery drivers can air up the tires on their cars and trucks. Install drops with outlets in all technical locations. They may not have a requirement right now, but you can bet your life they will in the future. People have a way of coming up with all kinds of uses for compressed air when it becomes available. Think about how you'll interconnect to the system. Hoses should be kept as small and as short as possible. Can you connect your hoses permanently or will you need quick disconnects? One thing to consider is that a permanent connection provides good security for your expensive air tools.

This chapter will take you through a series of step-by-step guidelines on specifying your compression system. When the time comes for you to go out for quotes, you won't be at the mercy of the salesmen.

Estimating Required SCFM

Determining your SCFM requirement is the most difficult task. After you get through this, it's all downhill. The problem is that if you simply add up the SCFM of all your air tools, you will come up with a compressor size that is grossly larger (and grossly more expensive) than you really need. This is because air tools are rarely used continuously. Additionally, an employee may use two or three air tools at their workstation, but they certainly don't use them at the same time.

The Actual SCFM requirement for an air tool is based on its duty cycle. The duty cycle is the amount of time that any given tool is on versus the amount of time it is off. As an example, if you have an air wrench that is specified at 14 SCFM and you turn it on for 1 minute every 10 minutes, then it is only using an average of 1.4 SCFM. This is only one-tenth the specified amount or a 10 percent duty cycle. That is to say that the wrench runs for 1 minute (or 10 percent of the time) and is off for 9 minutes (90 percent of the time). To determine the actual air requirement, multiply the tool's SCFM rating by 10 percent (or 0.1) (14 SCFM × 0.1 = 1.4 SCFM). If the tool runs for 5 minutes every 15 minutes then it has a 33 percent duty cycle. Five minutes on (33 percent of the time) and 10 minutes off (66 percent of the time). Actual air requirement in this case would be: 14 SCFM × 0.33 = 4.6 SCFM.

The key to determining your compressor size requirement is understanding the duty cycle of every tool that is used in the plant. This may seem like an almost impossible task, but it's not. The best way to accomplish this is to make a list of all the air tools, along with their SCFM rating that are used or will be used in the plant. After completing the list you can go tool by tool, estimate its duty cycle, and calculate its actual SCFM requirement. Then simply add up all the results and you will have a rough estimate for your compressor size. Typically, I am rather liberal with my duty cycle estimates and after I get my SCFM results, I will usually double the compressor size if I am going to specify a reciprocating unit. I will multiply my results by 1.5 if I am specifying a screw compressor. This is a pretty good way to anticipate future growth and it rarely results in a grossly oversized compressor.

Use Table 13-1 to list your tools, estimate your duty cycle, determine the tool's actual SCFM, and figure the compressor size.

It should be noted that if you have a very consistent air requirement, such as a piece of packaging equipment, amusement park ride, or printing press, then it is possible to size the compressor to closely match the load.

TABLE 13-1   SCFM Work Sheet

Tool Type	SCFM Rating	Duty Cycle	Actual SCFM
		Total SCFM	

Selecting a Compressor

Once you have determined your SCFM requirements, you can start to consider what type of compressor you'll need. If your requirements are less than 120 SCFM, then there are three basic choices: a single-stage reciprocating, a two-stage reciprocating, or a screw compressor. In this smaller size arena (50 hp and down), I generally recommend selecting a packaged unit. These compressors are complete, ready-to-install

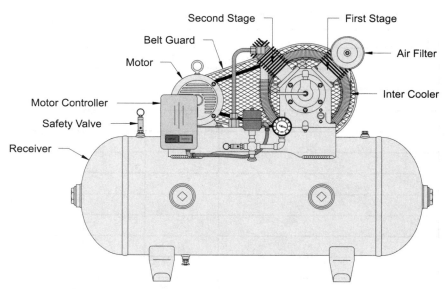

Figure 13-1   Two-stage reciprocating compressor.

systems that provide a lot of convenience when installing a compression system. For applications that have a lot of intermittent use, such as auto repair shops, reciprocating compressors seem to be the best choice. They offer a solid, proven solution that will provide many years of reliable service. If your application is less than 20 SCFM and intermittent, a single-stage compressor will provide the service that is required. For applications in the 30- to 120-SCFM range, a two-stage compressor as shown in Fig. 13-1 should be specified. For applications that are more steady state, a screw compressor may be a better solution. However, a reciprocating compressor can be set up for continuous-run applications and can be quite effective in this role. If a reciprocating compressor is chosen for a continuous-run application, then it should be equipped with an aftercooler. For requirements over 200 SCFM, a stand-alone screw compressor is the only real choice out there. Once you get to about 1000 SCFM (200 hp) you can start to consider turbo compressors, although they really don't become viable until about 2500 SCFM (500 hp).

Master Regulator

After selecting the compressor, you must consider the overall system pressure and how to achieve pressure control. For compressors under 40 hp, a master regulator (Fig. 13-2) should be used on the output of the

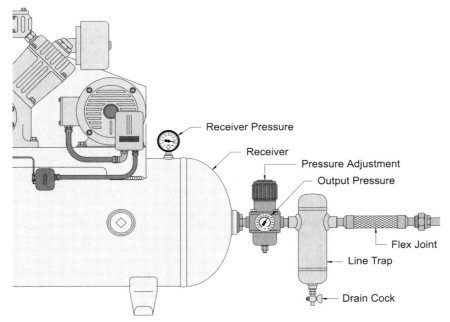

Figure 13-2 Master regulator.

primary receiver. For general purpose applications, set the output pressure to 90 psi. This is an adequate pressure for most compressed air tools. The flow rate of the regulator is also a consideration. A good rule of thumb is to size the regulator for a flow rate that is five times greater than the maximum rating of the compressor. As an example, a 20-hp compressor may have a maximum flow rating of 74 SCFM; therefore, the regulator should have a maximum flow rate of 370 SCFM. This oversizing of the master regulator is necessary in order to carry high-surge situations. For specific applications that require lower pressures, a point-of-use regulator should be used at the workstation or equipment connection. For systems over 40 hp, the compressor itself is usually set up to control the output pressure.

Receiver Size

Much to do is made over receiver sizing. If you do any research on the subject, you'll find all sorts of conflicting information. It's really tough to find a straightforward set of guidelines on how to select a receiver size. However, in the real world, if you're buying a compressor that is 30 hp or under, it will probably be delivered with an integral receiver as part

TABLE 13-2   Receiver Size Chart

Horse-power	Receiver Size (gal)	Receiver Size (cu ft)
1	10	1.34
2	20	2.67
3	30	4.01
4	40	5.36
5	60	8.02
7-1/2	80	10.70
10	120	16.04
15	120	16.04
20	200	26.74
25	200	26.74
30	240	32.09
40	400	53.48
50	500	66.84
75	1060	141.71
100	1550	207.22
200	2560	342.25
250	3000	401.07
350	3800	508.02
500	5000	668.45

of the package. For compressors over 30 hp and base-mounted units, Table 13-2 shows receiver sizes against compressor horsepower as appropriate for general purpose compressed air applications.

Figure 13-3 shows two different commercial receivers. The 120-gal unit is intended for use with packaged compressor designs, while the 240-gal unit is intended for stand-alone applications.

It should be noted that all receivers should be equipped with a pressure relief valve that is set to open at a pressure no higher than the maximum pressure rating on the receiver's name plate and a flow rate that is 20 percent higher than the maximum delivery rate.

Pipe Sizing

Pipe sizing is very important in any compression system. Many plants are plumbed based on guesswork, rather than the proper data. Invariably, this leads to a situation that pushes the limits of the delivery system and, in most cases, all manner of patches and band-aids are applied to try to work around the problem.

Be smart. Don't use old plumbing systems that may be left over from some other application. It is very difficult to determine the condition of the old pipe, fittings and valves, which in many cases, weren't designed for compressed air in the first place. Adapting an existing plumbing system to your compressed air requirements usually requires more work to bring it up to specification than simply putting in a new system.

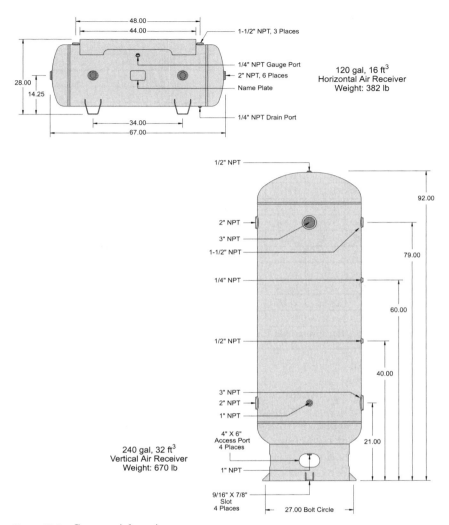

120 gal, 16 ft³
Horizontal Air Receiver
Weight: 382 lb

240 gal, 32 ft³
Vertical Air Receiver
Weight: 670 lb

Figure 13-3 Commercial receivers.

Follow the recommendations outlined in later pages and use the data provided in Table 13-3 to select the proper size pipe for your specific requirement.

Remote Receivers

Remote receivers are typically used to feed intermittent high-flow applications. It can be a little daunting determining the size requirements

TABLE 13-3   Recommended Pipe Sizes

		Length of Run in Feet								
		25	50	75	100	150	200	300	500	1000
	6	1/2	1/2	1/2	1/2	1/2	1/2	1/2	3/4	3/4
	18	1/2	1/2	1/2	3/4	3/4	3/4	3/4	1	1
	30	3/4	3/4	3/4	3/4	1	1	1	1-1/4	1-1/4
	45	3/4	3/4	1	1	1	1	1-1/4	1-1/4	1-1/4
Flow Rate in SCFM	60	3/4	1	1	1	1-1/4	1-1/4	1-1/4	1-1/2	1-1/2
	90	1	1	1-1/4	1-1/4	1-1/4	1-1/4	1-1/2	1-1/2	2
	120	1	1-1/4	1-1/4	1-1/4	1-1/2	1-1/2	1-1/2	2	2
	150	1-1/4	1-1/4	1-1/4	1-1/2	1-1/2	2	2	2	2-1/2
	180	1-1/4	1-1/2	1-1/2	1-1/2	2	2	2	2-1/2	2-1/2
	240	1-1/4	1-1/2	1-1/2	2	2	2	2-1/2	2-1/2	3
	300	1-1/2	2	2	2	2	2-1/2	2-1/2	3	3
	360	1-1/2	2	2	2	2-1/2	2-1/2	2-1/2	3	3
	450	2	2	2	2-1/2	2-1/2	3	3	3	3-1/2
	600	2	2-1/2	2-1/2	2-1/2	3	3	3	3-1/2	4
	750	2	2-1/2	2-1/2	3	3	3	3-1/2	3-1/2	4
	1000	2-1/2	3	3	3	3-1/2	3-1/2	4	4	5
		Recommended Pipe Sizes @ 100 psi (Schdule 40 Pipe)								

for one of these applications. There really isn't any cut-and-dry method for this problem. Each application must be assessed on its own merits. Determining the size of remote receivers is really out of the scope of the book. However, it is such an important aspect of some compression systems that I feel should be reviewed anyway. To size a remote receiver to feed an intermittent high-SCFM load, use the following formula:

$$[T \times (C - S) \times 14.7] \div (P_1 - P_2) = V$$

where V = receiver volume (ft^3)
    T = time (min)
    P_1 = start pressure
    P_2 = ending pressure
    C = intermittent flow requirement (SCFM)
    S = air supply flow (SCFM)

Suppose we have a compressor that is capable of delivering a flow of 80 SCFM at 90 psi. We must operate an air wrench that requires 96 SCFM for 1 minute and the delivery pressure can't drop below 65 psi. After the operation, the wrench is turned off for 30 seconds. What receiver size would we require to operate the wrench within the above parameters?

$$[1 \times (96 - 80) \times 14.7] \div (90 - 65) = 9.41 \text{ ft}^3 \text{ or } 70.3 \text{ gal}$$
$$(\text{standard size: 80 gal, } 10.70 \text{ ft}^3.)$$

Next we need to determine if the system can recover in the 30 second "off" period. Use the formula below to calculate recovery time:

$$[V \times (P_1 - P_2)] \div (14.7 \times S) = T$$

$$[10.70 \times (90 - 65)] \div (14.7 \times 80) = 0.23 \text{ min or } 13.8 \text{ s}$$

Note: All receivers should be equipped with a pressure relief valve that is set to open at a pressure no higher than the maximum pressure rating on the receiver's nameplate and a flow rate that is 20 percent higher than the maximum delivery rate.

Cost Savings

A common topic when discussing compressed air systems is cost savings. These days a great deal of emphasis is being placed on cost-saving equipment and devices. Take every suggestion for cost savings with a grain of salt. In most cases, replacing and/or adding new equipment or gadgets will net you no measurable savings whatsoever.

In real life, the number one loss associated with compression systems is air leaks. Yes, that's right, air leaks. Some sources estimate as much as 40 percent of all compressed air generated is lost to leaks or misuse!

Consider the case of a 50-hp compressor that is leaking 25 percent of its output. That means 12.5 hp of compressed air is continuously lost. At 1 kW/hp that's 100 kWh per 8 hour shift. If you figure two shifts per day, a 5-day workweek at 52 weeks a year, that's 52,000 kWh of wasted energy per year. If you're paying 8 cents per kWh, then that's an expense of $4160.00 per year!

In one instance I was called to quote a new dryer for a large machine shop. The company had two 50-hp screw compressors that ran continuously and one of the dryers was definitely past its prime. When I arrived, I was taken on a tour of the facility. I noticed the hissing of compressed air leaks everywhere I went. I suggested that before purchasing a new dryer, it would be a good idea to conduct an air audit. The report that was provided outlined every leak and misuse of compressed air in the plant. After my customer faithfully repaired and/or corrected every item on the list, they were able to completely shut down one of the compressors and place it into backup service. This provided them with a savings of $12,480.00 per year! On top of that they didn't have to buy a new dryer at about $5500.00. The air audit cost only $650.00, well worth it.

There are a number of locations on a compression system where leaks commonly occur. The compressor itself should be inspected yearly for developing leaks. Compressor leaks are not readily apparent because the noise of the pump overwhelms the noise of the leak. It seems like industrial quick disconnects and universal hose couplings are always leaking. The only thing you can do with the quick disconnects is replace them when necessary. Universal hose couplings should have their gaskets inspected and replaced on a regular basis. Hose barbs are another common leak source. Routinely check, tighten, and/or reterminate hose barb fittings as required. Swivel joints start to leak after a few years of service and should be routinely inspected and replaced as required. Be aware that some mechanics make a type of homemade swivel by loosely fitting two street elbows together. This not only creates a leak, but the swivel can accidentally unscrew and you have a whipping hose in your hands. This practice should be aggressively discouraged. Valve stems can easily start to leak after a few years of service. This is especially true of valves that see a great deal of use. In any case, every valve in a compression system should be regularly inspected and operated at least once every six months.

Drain valves can represent a serious problem area. Employees have a tendency to crack open drain valves to continuously discharge water. On any system, this is the most costly way to drain water. If you crack open five drains at 1 SCFM each, the associated cost would be $1950.00 per year! Installing automatic or self-closing drains will solve this problem. Carefully consider the cost versus benefit when purchasing zero loss drain valves. These valves rarely provide enough savings to offset their cost. On the flip side, float valves usually require so much maintenance that their low purchase price is completely negated. A good, electronically controlled solenoid valve is the best choice for this application.

Miscellaneous

Another area that just kills me whenever I see it is the use of an air jet as a fan. I have seen air lines just opened up to act as a fan for everything from cooling a piece of equipment to keeping a machinist comfortable. This is another practice that should be discouraged.

Any compressor will have ideal operating parameters. To get the most efficient use of your compressor, it should be carefully matched to your requirements and set up according to the manufacturer's recommendations. For systems that have a large flow variance, reciprocating compressors are good choices. If the system is too large for a reciprocating compressor, then a screw compressor with a variable frequency drive should be selected. Either of these types of compressors will match their energy consumption to a variable load.

Stripping out harmful water from the system will provide substantial long-term savings. Air tools that are fed water-free air will have a dramatically improved life. An air tool that gives you a two-year average service life will last 20 years if the air is free of water and the tool is properly lubricated. Take care of your tools and they will take care of you.

Employee training is another way to improve the efficiency of your compression system. Most users of compressed air don't know the first thing about air compression. Teach your people the dos and don'ts of compressed air. Make sure that they report any problems they encounter within the system. See to it that their air tools are properly inspected and serviced on a regular basis. A tool that is worn out may seem to be working OK, but it's taking your mechanic four times as long to do his job as it should. That's costing you money and your employees should be encouraged to point out these deficiencies.

Distribution Layouts

There are three different approaches to laying out compressed air distribution systems. The first and, unfortunately, the one we are forced to deal with most of the time is a collection of pipes and hoses that have been added as required, with little or no real thought whatsoever. Generally speaking, it is easier to simply replace a system like this rather than trying to salvage it. Be particularly cautious when adapting old piping systems. It's difficult to assess the internal condition of pipes and fittings and it's likely that the system wasn't even designed for compressed air in the first place.

The second approach is the use of a central spine that feeds the necessary drops. Figure 13-4 shows a typical spine distribution system.

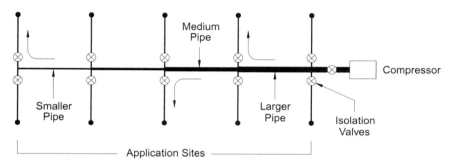

Figure 13-4   Spine distribution system.

This method of distribution is an excellent choice for smaller facilities. An auto repair garage, for instance, will typically have a spine running along the back wall of the shop with the drops extending out into the shop between the bays. Notice that the spine uses a larger pipe near the compressor and becomes smaller along its length as the flow requirements diminish. Also note the location and abundance of isolation valves. Applying an isolation valve to every drop will allow maintenance to be carried out in specific locations without disturbing operation of the overall compression system.

The third method is the loop distribution system as illustrated in Fig. 13-5. These systems are typically found in larger facilities. The system generally consists of a pipe loop that follows the outer wall of the building with drops taken off when needed. Loop systems also have the added benefit of flow equalizing. When a particular application is experiencing a surge situation that might affect the flow requirements of applications down the line, this system will automatically compensate by feeding air from the opposite side of the loop. Because of this situation the primary loop and any balance pipes can operate with bidirectional flow. This should be taken into consideration when selecting isolation valves. Only valves that will seal in both directions should be used on a loop distribution system. Balance pipes are placed at locations that will connect the opposite sides of the primary loop. Properly located balance pipes will help assure a consistent delivery rate to all application sites. The location and number of isolation valves is critical in a large distribution system. These valves allow maintenance to be conducted to any part of the system without disturbing the operation of the overall system. When installing two compressors, it is generally advantageous to place them on opposite sides of the loop, as shown.

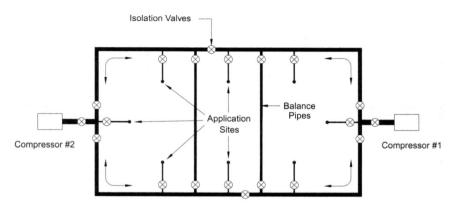

Figure 13-5   Loop distribution system.

Plumbing

More often than not, the plumbing which is installed to service any given compression system is wholly inadequate. There are a few basic guidelines to follow when installing or adapting the plant compressed air piping. Figure 13-6 should be used as a guide to designing, modifying, or evaluating compressed air distribution systems. Traps are very important to any compression system. Similarly, the method used to connect drops to overhead pipes is very important. Valves should be liberally applied, as these will provide easy servicing to individual sections of the system. Take particular notice of the rise in the horizontal pipes, as this allows water to freely drain back into the traps.

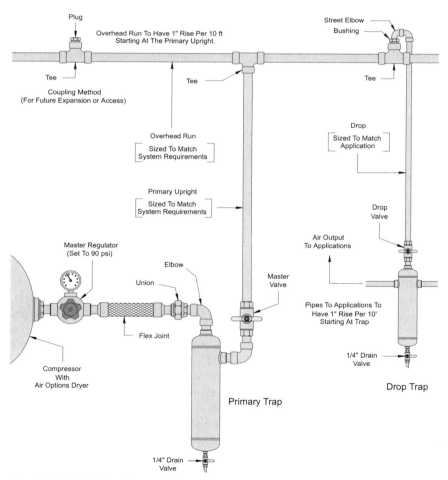

Figure 13-6   Plant plumbing.

CAUTION: Under no circumstances should PVC pipe or fittings be used for compressed air! PVC pipe is not intended for compressed air applications because of the inherent stored energy of the gas. It is only intended for non-compressible fluids. The reason for this is that PVC pipe normally fractures when it fails. As the energy of the compressed air is released, a shock wave travels down the joint of pipe until it encounters a strong point, usually a fitting. As the shock wave propagates along the pipe, it continuously fractures and blasts out hundreds of razor sharp, high-speed shards of plastic. I have personally experienced this phenomenon and can say with certainty that it is not funny and it creates a severe safety hazard. To make matters worse, PVC can react with some compressor oils and form microsurface cracks on the inside of the pipe and fittings. This, in turn, weakens the pipe and creates an almost perfect scenario for this type of failure.

Large compressed air distribution systems (Fig. 13-7) have their unique challenges. For very long overhead runs, it may be impractical to provide a 1 in. rise per 10 ft through the entire run. In these cases a short drop is fed into a drop trap at a lower level and a secondary run is continued from the output of the trap. The trap's drain is plumbed down to floor level.

For applications that require high demands for short periods of time, additional receivers may be placed through the system. The receivers allow high peak loads to be supplied by a rather modest overhead run. In most instances, sizing the overhead run to feed a high peak load would require prohibitively large piping. It should be noted that caution must be exercised when installing auxiliary receivers. Their added volume and use may increase the load on the compressor, and too many may create a situation that could pull the air pressure down to unacceptable levels. Refer back to section "Remote Receivers" for information on sizing a remote receiver for your particular application.

Buried Pipe

It is always preferred to run compressed air lines overhead. However, there are situations that demand that underground lines be installed. The problem with underground lines is that they form an almost perfect water trap and once the pipe fills with water, it's almost impossible to drain. Figure 13-8 shows a basic layout for underground compressed air piping. The line traps on either end of the run are critical. A 1 in. rise per 10 ft of run, starting at the compressor end, is also recommended. This allows water to flow freely back into the line trap so that it may be periodically drained. When sizing the pipe, consider future expansion. Nothing is quite so irritating as ripping up 100 ft of driveway to replace a pipe just because it isn't big enough. On the rare occasions that compressed air piping must be buried, it's a good idea to install a pipe that is twice as large as the requirement calls for. Additionally, all buried pipe and fittings should be coated for corrosion protection.

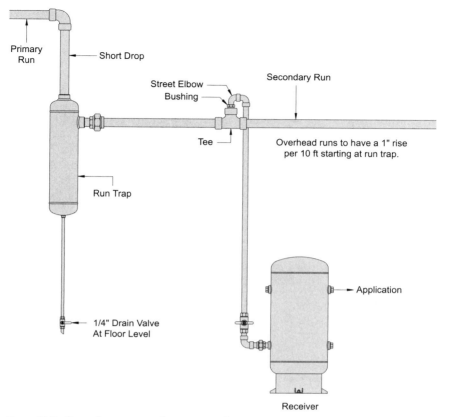

Figure 13-7 Secondary runs and remote receivers.

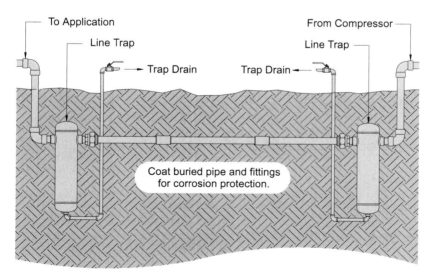

Figure 13-8 Underground piping.

Plumbing Tools

Before starting a compressed air distribution project, it's a good idea to familiarize and equip yourself with a few of the more common plumbing tools. Having a set of quality plumbing tools, which are in good condition, will make the job go a lot easier. Figure 13-9 shows the hand tools that are normally used when installing or modifying a distribution system.

The wire brush is used to clean pipe threads before assembly. A 25- to 50-ft tape measures for measuring pipe and building lengths. A 24-in. magnetic base level is extremely handy. The magnet allows the level to stick to the pipe being leveled, freeing the plumber's hands for other

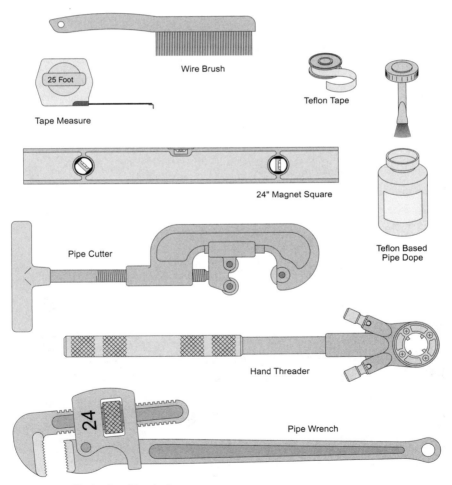

Figure 13-9 Basic plumbing tools.

tasks. A solid pipe cutter is a must. Be sure that the cutter wheel is sharp and free from any chips. Do not confuse tubing cutters with pipe cutters. Although tubing cutters have the same general appearance as a pipe cutter, they are considerably lighter in weight and will not stand up to the heavier loads required to cut pipe. A hand-operated pipe threading set is also a must. A good threading set will have dies for $1/8$ to 2-in. pipe. It should be noted, however, that hand-threading pipe over 1 in. requires a great deal of effort and is best left to powered threading machines. Pipe wrenches are a pivotal tool when conducting a plumbing project. For any given toolbox, there should be a 10-, 12-, 24-, and 36-in. pipe wrench, as shown in the illustration. Be certain that you keep the jaws sharp. This will ensure that the wrench bites properly and doesn't slip when loaded. For 24- and 36-in. wrenches it is recommended that aluminum units are selected. Forged steel wrenches in this size range are rather heavy and can be difficult to use.

All pipe joints should have a sealant applied before assembly. The best sealant is a Teflon-based pipe dope. Although it doesn't provide as good of a seal, for small repair jobs Teflon tape is often used. The Teflon component of the dope and tape provides the necessary lubrication so that the threads can seal properly. As the threads seat, the pipe dope or tape is forced into the remaining gaps and forms an air-tight seal.

A pipe vise can make even a small project a lot easier. Figure 13-10 shows a typical bench-mounted pipe vise. The unit can be hinged open

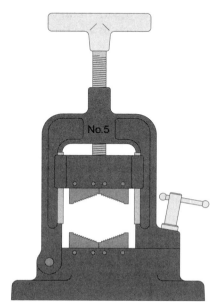

Figure 13-10   Pipe vise.

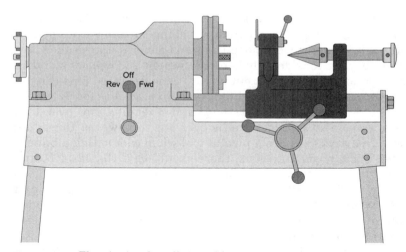

Figure 13-11   Electric pipe threading machine.

to accept pipe and will provide exceptional clamping force. For large plants, these vises are often mounted on the rear bumper of a truck or a portable tripod bench.

Figure 13-11 shows a typical powered pipe-threading machine. These machines are necessary for all but the smallest distribution projects. A typical pipe-threading machine will have dies for threading $^1/_8$ to 2 in. and includes an integral pipe cutter with inner diameter reamer. These machines will have reversing motors and can be used with hand threaders. Most units are equipped with a foot switch, a cutting oil pump, and wheel-barrow-type wheel set. Most powered pipe threaders can also be equipped with dies for threading bolts as well as pipe.

Questions*

1. What is the first consideration when specifying a compression system?

   (A) Space   (B) Power requirements   (C) Its use   (D) Location

2. In addition to the air supply, permanently connected hoses provide what?

   (A) Good flow   (B) Security   (C) Longer life   (D) Lower costs

3. Which is the most difficult task when specifying a compression system?

   (A) Power requirements   (B) Tank size   (C) Pipe size
   (D) SCFM requirements

4. When determining the SCFM for a tool what must you consider?

   (A) Duty cycle   (B) Actual SCFM   (C) Oil requirements
   (D) Hose size

5. If a tool runs for 1 minute and is off for 9 minutes what is its duty cycle?

   (A) 10 percent   (B) 90 percent   (C) 50 percent   (D) 100 percent

6. Knowing your Actual SCFM, what multiplier should be used when specifying the compressor?

   (A) 5×   (B) 1.5× for screw   (C) 2× for reciprocating   (D) 10×

7. Compressors below 30 hp are generally configured as what?

   (A) Indoor   (C) Outdoor   (D) Packaged   (C) Skid mounted

8. What type of compressor should be selected for an intermittent application in the 30- to 120-SCFM range?

   (A) Screw   (B) Two-stage, packaged   (C) Single stage
   (D) Liquid ring

9. What type of compressor should be specified for a steady state application in the 200- to 500-SCFM range?

   (A) Turbo   (B) Screw   (C) Reciprocating   (D) Diaphragm

10. What is a good rule of thumb multiplier for sizing a master regulator?

   (A) 2×   (B) 10×   (C) 8×   (D) 5×

11. What receiver size should be selected for a 100-hp compressor?

   (A) 120 gal   (B) 1550 gal   (C) 25 gal   (D) 5000 gal

---

*Circle all that apply.

12. What essential piece of equipment should all receivers be equipped with?

(A) Drain valve   (B) Pressure gauge   (C) Output valve
(D) Safety valve

13. What pipe size should be used for a run that is 200-ft long and is supplying 100-psi air at 150 SCFM?

(A) 1 in.   (B) 2 in.   (C) 4 in.   (D) 12 in.

14. What is the number one loss associated with compressed air?

(A) Oil consumption   (B) Water   (C) Paint   (D) Leaks

15. A small plant will generally use what type of distribution system?

(A) Spine   (B) Hose   (C) Loop   (D) Temporary

16. A large plant will generally use what type of distribution system?

(A) Spine   (B) Hose   (C) Loop   (D) Temporary

17. What type of pipe should never be used for compressed air?

(A) Steel   (B) Stainless steel   (C) Copper   (D) PVC

18. It is always better to run pipe where?

(A) Overhead   (B) Underground   (C) On the wall   (D) Outside

19. What is the best pipe thread sealant?

(A) Glue   (B) Pipe dope   (C) Teflon tape   (D) Tar

14

Conducting an Air Audit

An air audit is simply a detailed inspection of a compressed air system. The principal reason for conducting an air audit is to generate a clear understanding of the compression system that you are working on and in particular provide a detailed list of all the deficiencies that the system may have. Compression systems are rarely designed in advance. In the case of smaller systems, the compressor is usually purchased first and then everything else is considered after the fact. Oftentimes, a preexisting system is placed into service with little or no thought as to how suitable it might be for the specific application.

To make matters worse, compression systems are normally evolutionary in nature. Over the years, compressors will be resized and changed, dryers and aftercoolers added or removed, processes added, locations changed, and whole new sections of distribution piggybacked onto the existing network. For the most part, compression systems grow and rarely, if ever, will a plant engineer be called in to consider the impact on the overall system, when making these changes. Additionally, we humans definitely have an *"if it's not broke, don't fix it"* attitude, which allows compression systems to slowly grow and deteriorate over many years. This translates to millions of dollars in hidden costs every single year within the compression industry.

In one instance, in particular, I was called to inspect a dryer and make a recommendation for repair or replacement. The system was fed by a single 500-hp turbo compressor. When I arrived on site, it was immediately apparent that there was a significant air leak in the compressor. It seems that the pressure control system had failed and the head of maintenance decided that he didn't want to spend the money to replace it. So instead, he added a gate valve to the output of the compressor and simply bled off air until the discharge was down to the system pressure they required. A quick calculation showed that they were continuously dumping about 900 SCFM of air! The head of maintenance commented that it didn't matter because the compressor was running anyway. When I pointed out the flaw in his argument and described how the compressor was supposed to operate, I was curtly told that my job was to review the dryer and I had no business commenting on any other aspect of the system. I completed my dryer recommendations and left. This is a clear case of an individual who does not have the necessary knowledge to make important decisions concerning the compression system that had been placed in his care. As it was, the company's cost to operate the system is greatly inflated and will most likely increase as the years wear on.

One thing to bear in mind is that if you hear air escaping, it is costing money. Leaks are the number one loss associated with compressed air systems. A quick method to gauge the soundness of the distribution system is to walk the plant after it has been shut down. Turn on the compressor and allow the system to come up to full pressure. While all other equipment is shut down, walk the compressed air paths and make a note of any hissing that you might hear. Another method is to bring the system up to full pressure and then turn off the compressor. Time how long it takes for the system pressure to bleed down to zero. Ideally, the pressure shouldn't drop at all. I have been told by some rather knowledgeable people that it is impossible to create a leak-free compressed air system. This is simply hogwash. If you can have a leak-free water system, then you can have a leak-free air system. The only real difference in the two is that water leaks make a big mess and air leaks don't. So guess what, the water leaks get fixed and the air leaks don't. And you know what else is interesting—an air leak costs you more money than the same size water leak does.

One other thing that is worth mentioning: don't be surprised if you get a lot of resistance from the maintenance guys. An air audit with poor grades is generally a clear indication that there is a systemic problem within the maintenance department.

When conducting an air audit, it is critical that you thoroughly evaluate every aspect of the compression system. Although poorly performing compression systems will generally have one or two significant problems, usually the biggest drag on the system comes in the form of many small problems. A slipshod inspection of the system may uncover the big problems, but could miss the bulk of the small problems and very little improvement may be realized. I have found that a systematic approach is the best way to go about this process. I generally use an outline as a guide and check off items after I have conducted the task and made all pertinent notes.

Distribution Map

The next step is to generate a detailed distribution map. This is generally an engineering drawing of the system as it exists. Most compressed air distribution systems, especially in small companies, are designed and installed by a plumber or maintenance man with little or no knowledge of air compression. Some companies may have an outdated drawing or sketch that can provide a good start. However, do not trust company supplied drawings. Verify every aspect of the drawing before using it. Air systems are routinely modified on the spot and these modifications are usually not reported to the engineering group, if one exists at all. I will normally make sketches of the system with pipe sizes, fittings, rise and fall measurements, drop locations, quick disconnects, valves, and the like. I use these detailed sketches to generate a complete engineering drawing of the system. I usually use a CAD system to make the drawing. The drawing may be used as a base for all the other inspections that will be made during the audit. As each item is completed, the drawing may be updated to reflect added information. Figure 14-1 shows a sample distribution map drawing.

SCFM Requirements

After the distribution map is complete, it seems that the best place to start an air audit is at the user level. Make a list of all compressed air users, estimate their SCFM requirement, and make notations on the distribution map. Inspect the condition of all the tools or components and make recommendations for rebuild or replacement. Some tools can be sent for testing and some must be evaluated on site. One method that I have used successfully is to interview the employees that use the tools. You can be sure that the shop guys know what grinders work best and which are junk. Don't be shocked at how many problems you might find. It is not uncommon that recommendations are made to repair or replace as much as 90 percent of all tools and fittings within a company.

Leaks

The next step in the process is to inspect for leaks. This is a very important aspect of an air audit. There are two principal types of leaks, ones that can be heard and ones that can't. The noisy ones are easy to find. However, in a noisy manufacturing plant, even gross leaks may not be easy to hear. A good way to handle this problem is to conduct this inspection when the plant is shut down. Make sure all plant equipment is turned off, including all air-handling equipment, that is, air conditioners,

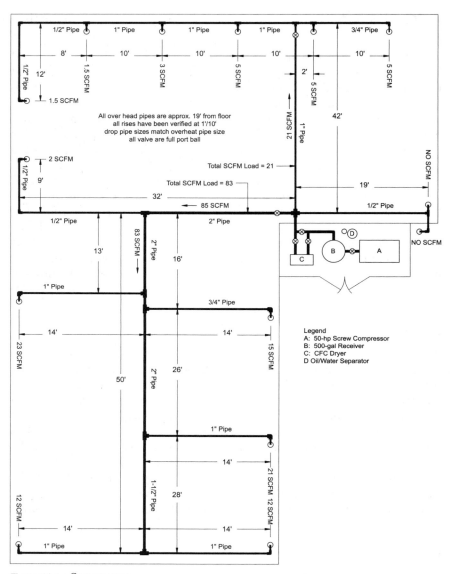

Figure 14-1 System map.

heaters, fans, and the like. Turn on the compressor and allow the system to come up to full pressure. While all other equipment is shut down, walk the compressed air paths and make detailed notes of any leaks that you detect. The smaller leaks are a little more difficult to find. But, there is a trick that can be used that will speed up the process. Inspect every fitting, joint, and valve stem in the system. Any of these locations that are leaking will seem wet. This is the water and oil vapor that condenses onto the fitting as the leaking air expands. Check the suspect joint or seal with soap bubbles to verify a leak. As leaks are identified, the drawing should be marked and a yellow tag should be wired to each and every leak site. In the event that you identify a leak in the pipe wall itself, this means that the pipe has excessive internal corrosion and the pipe should be immediately taken out of service and replaced. Other sections of pipe within that run should be carefully inspected and replaced as necessary. When the first pinhole leak occurs, it generally means that the rest of the pipes are not far behind. Once again, don't be shocked if, after spending the entire night tagging leaks, you look back on the plant and see an ocean of yellow tags. Remember, leaks are the number one loss associated with compressed air systems and that means there are a lot of them that need to be fixed.

Hoses

Hoses are another problem area. Hoses are oftentimes neglected until they actually rupture. All hoses should be inspected for leaks, abrasion, age deterioration, size, and application. It is very common for employees to expand a compressed air system by simply installing a semipermanent hose to an application. I have even seen some companies who have constructed their entire compressed air distribution system out of hoses alone! This practice should be aggressively discouraged. Inspect all hose connectors and hose terminations. A list of all hose and connector problems should be compiled and locations noted on the distribution map.

Filters

Most compressed air systems don't even have filters. It simply doesn't occur to compressed air users that filters may benefit their operation. Recommendations should be made for filters and their locations noted on the distribution map. The recommendations should include filter size, micron rating, flow rates, and service intervals.

Regulators

All regulators, including the master regulator, should be inspected for their overall condition, pressure settings, and flow capabilities. A recommendation should be made to replace any regulator that is not in good, serviceable condition. Additionally, it is commonplace to operate air tools at pressures higher than the manufacturer's maximum rating. In these cases, a recommendation should be made to install a line regulator. All regulator locations and additions should be noted on the distribution map.

Lubricators

Lubrication is a highly neglected aspect of compressed air systems. If lubricators are not used within the system, then a review of lubricating procedures and intervals should be conducted. Generally speaking, there is a complete absence of any lubricating procedure and lubrication is left entirely up to the users of the tools. This means that the tools never get lubricated. Each application should be evaluated and a recommendation should be made to install a lubricator as necessary. Recommendations should include size, oil type, delivery rate, and service intervals. Additionally, any lubricator that is not in serviceable condition should be replaced. At sites that require clean air as well as lubricated air, it is necessary to install dual outlets. All lubricator locations should be noted on the distribution map.

Remote Receivers

Remote receivers are found on many compression systems. All receivers within the system should be inspected and qualified by a licensed and bonded tank inspection service. If any receiver is found to be missing its original specification nameplate then that receiver should be taken out of service and destroyed or requalified by the tank inspection service. Particular attention should be paid to the tank's external paint. A deteriorating paint job will allow rust to take hold in the confined areas around the mounts and nameplate. The rust will slowly work its way through the receiver wall and leaks will develop. In this case, the only recourse is to replace the receiver. The inside of the tank should undergo a detailed visual inspection and any problem areas should be noted and qualified. Finally, the tank may require hydrostatic testing to requalify its pressure rating. It should be noted that the type of inspection and testing described can be rather expensive. In the case of small receivers, it may be more cost effective to simply replace any tank that is suspect.

The size of remote receivers should be verified following the sizing procedure outlined in Chap. 13, section "Sizing Remote Receivers." Verify that all receivers within the system are equipped with a pressure relief valve that is set to open at a pressure no higher than the maximum pressure rating on the receiver's nameplate and have a flow rate that is 20 percent higher than the maximum delivery rate to the receiver in which they serve. Carefully inspect the pressure relief valves and recommend replacement of any valve that is suspect. Inspect the drain valves on the bottom of the receiver for leaks and general condition.

Take note of the amount of water present in the receiver when you inspect it. If there is a substantial amount of water inside the receiver, then this indicates that it is not being drained often enough and an automatic drain should be added. Recommendations should be made for adding, removing, and replacing remote receivers and their support equipment. The remote receiver's particulars and locations should be noted on the distribution map.

Master Regulator

The master regulator is an exceptionally important component within the compression system. It should be carefully inspected for its overall condition, pressure setting, and flow capability. The regulation of the unit should be evaluated by monitoring the output pressure during normal plant operations. If the output pressure changes any more than ±5 psi, the regulator is not functioning properly and should be replaced. It should also be noted that lower-cost regulators usually have rather poor output regulation, which may wander as much as ±20 psi! If a low-cost regulator is being used, then a recommendation should be made to replace it with a higher quality unit.

Primary Receiver

A primary receiver is incorporated into most compression systems. It should be inspected and qualified by a licensed and bonded tank inspection service. If the receiver is found to be missing its original specification nameplate, then it should be taken out of service and destroyed or requalified by the tank inspection service. As with remote receivers, particular attention should be paid to the tank's external paint. A deteriorating paint job will allow rust to take hold in the confined areas around the mounts and tag. The rust will slowly work its way through the receiver wall and leaks will develop. In this case, the only recourse is to replace the receiver. The inside of the tank should undergo a detailed visual inspection and any problem areas should be noted and qualified. Finally, the tank may require hydrostatic testing to requalify its pressure rating. As with remote receivers, the type of inspection and testing described can be rather expensive. In the case of small receivers, it may be more cost effective to simply replace any tank that is suspect.

The size of the primary receiver should be verified against the Table 13-2. Verify that the receiver is equipped with a safety relief valve that is set to open at a pressure no higher than the maximum pressure rating on the receiver's nameplate. Carefully inspect the pressure relief valve and recommend replacement if the valve is suspect. Inspect the drain valve on the bottom of the receiver for leaks and general condition.

Take note of the amount of water present in the receiver when you inspect it. If there is a substantial amount of water inside the receiver, then this indicates that it is not being drained often enough and an automatic drain should be added. Recommendations should be made for adding, removing, or replacing the primary receiver and its support equipment. The primary receiver's particulars and location should be noted on the distribution map.

Dryers

Determine what level of dry air is required within the system. Inspect the dryer(s) to verify that they are functional. It is not unusual to find dryers that are completely nonfunctional and their users don't even know it. The best way to evaluate dryer performance is to place a dew point analyzer downstream from the dryer. If a dew point analysis is not desirable, or deemed inappropriate, then simply check the system drains for water buildup. This is another area in which employee interviews can be rather telling. If the dryers are not performing up to specification then the problem must be tracked down. Check input temperatures and verify that they are within the manufacturer's recommendations. The maximum input temperature and SCFM rating is normally printed on the dryer's nameplate. In the case of refrigerated dryers, inspect the condenser coil for dirt. Also inspect the aftercooler on the compressor. It is not uncommon to find aftercoolers and condenser coils that are completely clogged with dirt, which forces the dryer to operate within unacceptable parameters. After cleaning the coils, verify that the gauges on the dryer are within their green bands. If they are not, then the dryer must be serviced or replaced.

In the case of desiccant dryers, the condition of the desiccant is always suspect. If the desiccant dryer does not have a first-stage dryer as discussed in Chap. 6, a recommendation should be made to add one. If the desiccant is oil contaminated, it must be replaced. If the desiccant charge has been in service for more than 2 years, it is probably a good idea to recommend replacement. After the dryer(s) are brought up to specification, a recommendation should be made to regularly service the unit(s) per manufacturer's guidelines.

Compressors

Verify that the compressor has an SCFM rating that will support the system's flow requirements. Review the compressor installation. Is it appropriate for the particular application, that is, duplex, redundancy, peak demand, reciprocating, screw, and the like? If the installation or capacity is inappropriate for the system, then a recommendation must be made to upgrade, expand, or replace the compressor(s). Check the entire compressor installation for leaks. Use caution, some components on an operating compressor become very hot and will cause severe burns if you come in contact with them. Inspect the drive belts. They should be in good condition and taut. Exercise caution around the belts as they can be extremely dangerous when the compressor is in operation. Be certain that all belt guards are in place and properly secured. Inspect the electrical controls, connections, and disconnects. Use caution, if the compressor is in a poor state of repair, lethal voltages may be exposed. A recommendation for all safety repairs should be made at once and carried out by an authorized factory technician.

If the compressor(s) are found to be adequate, then a review of the service procedures and records should be conducted. In the absence of any meaningful service, the compressor(s) should undergo a complete inspection and servicing by a factory authorized service technician and a regular preventive maintenance program should be implemented. For smaller compressors, this program may be carried out by the company's maintenance department. For companies that do not wish to service their compressor and for larger compressors, a service agreement can be purchased through your local authorized service center.

Routine Maintenance

It goes without saying that a compression system is an extremely complex network that requires regular maintenance. A detailed preventive maintenance and inspection program should be designed and implemented. The program should include required service intervals, stock materials, inspection procedures, evaluations, and service logs. The program should be carried out by authorized personnel only.

Employee Training

One area that is almost always neglected is employee training. There are typically two different categories in this regard. The first is the training of maintenance personnel. They should be thoroughly trained to carry out the maintenance program designed for the system that has been placed under their care. If a factory service program is not purchased for the compressors and dryers, then the in-house personnel should either attend training classes at the manufacturer's location or review and understand all maintenance materials that are provided with the equipment. The maintenance personnel should be encouraged to make suggestions for improving the compression system, after all they are the ones who are working with it everyday. However, they should be discouraged from making any significant changes without first having a discussion as to how those changes may impact the overall system.

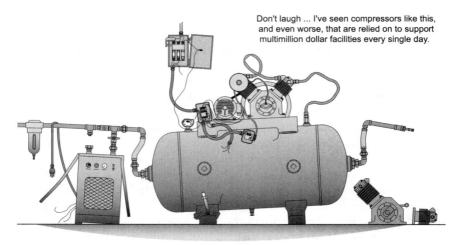

Don't laugh ... I've seen compressors like this, and even worse, that are relied on to support multimillion dollar facilities every single day.

Figure 14-2   Poorly maintained compressor graphic.

The second category is user training. This doesn't require a particularly complicated training program. Usually, a single sheet of guidelines that users are required to follow when working with air tools is sufficient. Users should also be encouraged to point out any problems that they notice within the system. They should also be discouraged from making any modifications to the system without express permission from their supervisor.

Sample Outline for Conducting an Air Audit

1. System map drawing
    1.1 Pipe sizes
    1.2 Pipe lengths
    1.3 Fittings
    1.4 Valves
    1.5 Drop locations
    1.6 Trap locations
    1.7 Rise data
    1.8 Primary receiver
    1.9 Remote receivers
    1.10 Filters
    1.11 Lubricators
    1.12 Regulators
2. SCFM usage
    2.1 Estimated SCFM work sheet(s) [Table 13-1]

3. Tools and equipment

    3.1 Inspection
    3.2 Performance test
    3.3 Employee interviews
    3.4 Rebuild or replacement
    3.5 Service intervals

4. Leaks

    4.1 Locations
    4.2 Corrections

5. Hoses

    5.1 Hose inspection
    5.2 Connector inspection
    5.3 Repair and replacement
    5.4 Service intervals

6. Filters

    6.1 Inspection
    6.2 Addition
    6.3 Filtration level

7. Regulators

    7.1 Inspection
    7.2 Size
    7.3 Replacement

8. Lubricators

    8.1 Inspection
    8.2 Delivery rate
    8.3 Replacement
    8.4 Addition
    8.5 Service intervals

9. Remote receivers

    9.1 Inspection
    9.2 Evaluate actual requirement
    9.3 Safety inspection
    9.4 Removal/replacement
    9.5 Draining

10. Master regulator

    10.1 Inspection
    10.2 Output pressure setting (90 psi)
    10.3 Replacement

11. Primary receiver

    11.1 Inspection
    11.2 Evaluate actual requirement

11.3 Safety inspection

11.4 Removal/replacement

11.5 Draining

12. Dryer

12.1 Inspection

12.2 Performance evaluation

12.3 Ambient parameters

12.4 Service per manufacturers guidelines

13. Compressor

13.1 Inspection

13.2 Performance evaluation

13.3 Ambient parameters

13.4 Service per manufacturers recommendation

13.5 Compare output to required load

14. Maintenance

14.1 Maintenance program

14.2 Maintenance log

15. Employee training

15.1 Compressed air usage

15.2 System maintenance

Questions*

1. What is an air audit?

    (A) A price review of the compression system.
    (B) A detailed inspection of the compression system.
    (C) A review of the costs associated with the system.
    (D) An inspection of the compressor.

2. Which group may resist the idea of an air audit?

    (A) Accounting   (B) Fabrication   (C) Shipping and receiving
    (D) Maintenance

3. Which is one of the most important items to look for during an air audit?

    (A) Leaks   (B) Heat   (C) Oil slicks   (D) Rusty pipes

4. Which is one of the most important documents for an air audit?

    (A) Lists   (B) System map   (C) Check list   (D) Outline

5. Which items should be inspected during an air audit?

    (A) Hoses   (B) Regulators   (C) Filters   (D) Compressor

6. After the air audit is complete, what should be implemented?

    (A) A regular maintenance program
    (B) Lunch
    (C) All repairs
    (D) Employee training

---

*Circle all that apply.

15

Single-Stage Home-Built Air Compressor

The following plans are for a small compressor intended for the home-garage shop. It is designed to be built very inexpensively from readily available parts that may be scrounged from many sources. The pump is made by converting an ordinary single-cylinder edger engine. The tank is a propane cylinder. The base and motor mount are constructed from a pine 2×10. All the plumbing is constructed from common fittings and the electrical is rather simple. The motor can be from $^1/_2$ to $1 \, ^1/_2$ hp and either 1725 or 3450 RPM. The design calls for 115 VAC operation but could be easily configured for 220 VAC. The maximum operating pressure is 60 psi and the CFM rating, depending on the motor used, is from 2 to 6 CFM. The finished compressor measures approximately 36 tall $\times$ 24 wide $\times$ 18-in. deep and weighs approximately 65 lb.

Figure 15-1 shows a general view of the finished compressor. Figure 15-2 shows the construction of the base. Figure 15-3 shows how the tank is prepared for mounting to the base and the tabs for mounting the motor mount. Figure 15-4 shows how to prepare the mounting tabs. Figure 15-5 shows the simple electrical diagram. Figure 15-6 shows the pickup tube assembly. A complete bill of materials and a written description for the preparation of the engine for pump service is given in Table 15-1. Written assembly instructions are outlined in the section "Assembly Instructions." Persons undertaking this project should take full advantage of the many sources for components such as: pawn shops, second hand stores, junk yards, surplus centers, thrift stores, garage sales, flea markets, friends, family, and neighbors. If you are patient and diligent in your collecting, you can build this compressor for very little money.

Air Compressor Bill of Materials

Item	Description	Qty.
1.	20 lb propane tank	1
2.	$1/2 \times 3/4$ NPT bushing	1
3.	$1/2$ NPT cross	2
4.	$1/2 \times 1/4$ NPT bushing	4
5.	$1/4$ NPT ball valve	1
6.	$1/4$ NPT pressure relief valve 150 psi	1
7.	$1/2$ NPT $\times$ 8″ pipe nipple	1
8.	$1/2$ NPT tee	3
9.	$1/4$ Pressure switch with unloader	1
10.	$1/4$ OD $\times$ 20″ copper tube	1
11.	$1/2$ NPT close nipple	3
12.	$1/2$ NPT coupling	1
13.	$1/4$ NPT coupling	1
14.	$1/4$ NPT, 2″, 0- to 100-psi pressure gauge	1
15.	$1/4$ NPT air quick disconnect	1
16.	$1/2$ NPT $\times$ $1/2$ compression $\times$ 90°	1
17.	$1/8$ NPT $\times$ $1/4$ compression $\times$ 90°	1
18.	$3/8$ NPT $\times$ $1/2$ compression fitting	1
19.	$1/2$ OD $\times$ 28″ copper tube	1
20.	$1/2$ NPT plug	2
21.	Pump sheave	1
22.	Motor sheave	1
23.	Vee belt	1
24.	$5/6 \times 2.50$ lag bolts	4
25.	$5/16 \times 2.00$ lag bolts	4
26.	$5/16$ flat washer	4
27.	$1/4$-20 $\times$ 1.00 hex bolt	8
28.	$1/4$-20 hex nut	8
29.	$1/4$ flat washer	20
30.	14-3 AC cord, 115 VAC	1
31.	Yellow wire nuts	2
32.	14-3 SJO	1
33.	$1/2$ NPT strain relief for 14 AWG romex	3
34.	$3/8$″ $\times$ close nipple	4
35.	10-32 $\times$ 1″ screw	1
36.	#10 flat washer	2
37.	4 place handy box with switch cover	1
38.	20 amp switch	1
39.	$2 \times 10 \times 18$	2
40.	$2 \times 4 \times 18$	2
41.	#10 $\times$ 3″ dry wall screw	10
42.	Compressor pump	1
43.	$3/4$ hp, 1725 RPM, 115-VAC motor	1
44.	$3/8$″ NPT check valve	2
45.	$1/2 \times 3/8$ NPT bushing	2
46.	1.00 $\times$ 1.50 $\times$ .12 angle $\times$ 1.00 long	8
47.	10-24 hex nut	1
48.	Engine gasket set	1
49.	Thin head gasket	1
50.	#10 $\times$ 1″ drywall screw	2

Air Compressor Bill of Materials (*Continued*)

Item	Description	Qty.
51.	$1/4$ NPT × 6″ nipple	1
52.	$1/4$ NPT tee	1
53.	$1/4$ NPT close nipple	1
54.	10-32 hex nut	1
55.	$1/4$ NPT × 24″ nipple	1
56.	10-24 × 4″ hex bolt	1
57.	#10 flat washer	1
58.	Teflon tape	1

Engine Preparation Instructions

Engine Type: 2.5 hp- to 5-hp Horizontal Shaft

Step	Description
1.	Completely disassemble engine.
2.	Discard: ignition coil, point set and associated parts, spark plug, carburetor and gas tank assembly (retain air filter), muffler, cam shaft and tappets and govenor.
3.	Completely clean all remaining engine parts.
4.	Remove the internals of the recoil starter and drive and discard.
5.	Inspect all parts including: bearings, seals, cylinder (to be free of any deep scoring), piston, and piston rings.
6.	Drill and tap spark plug hole to $1/2$ NPT.
7.	Lap valves and seats.
8.	Reassemble remaining engine components, less head, and flywheel cover.
9.	Measure bore and stroke in inches.
10.	Reassemble remaining engine components.
11.	Drill a $1/4$″ hole through the center of item 20, 1 only.
12.	Assemble items 8, 17, 18, 20, 34, 44, 45, and 58 per Fig. 15-1.
13.	Install assembly from step 12 on engine per Fig. 15-1.
14.	Install air filter and items 47, 56, and 57 per Fig. 15-1.
15.	Fill crankcase with 10 liter compressor oil. (Do not use motor oil!)

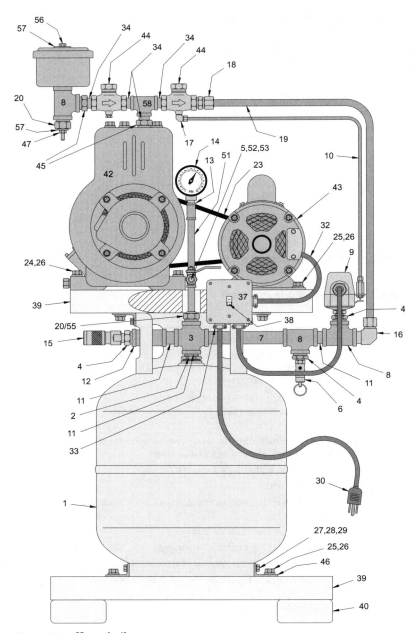

Figure 15-1 Home-built compressor.

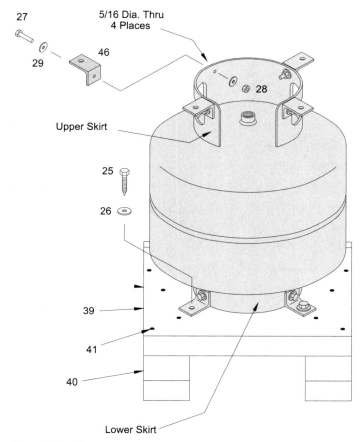

27

5/16 Dia. Thru
4 Places

29

46

28

Upper Skirt

25

26

39

41

40

Lower Skirt

Figure 15-2   Base.

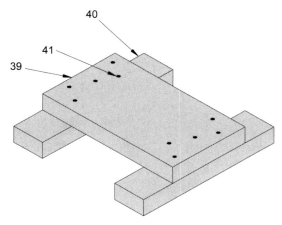

40

41

39

Figure 15-3   Tank assembly.

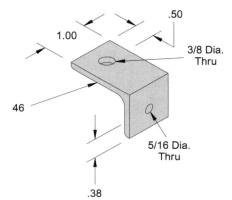

Figure 15-4   Mount bracket.

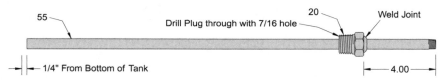

Figure 15-5   Electrical system.

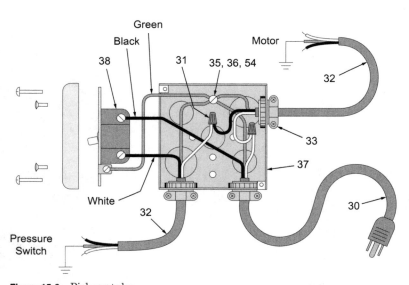

Figure 15-6   Pick-up tube.

Selecting the Sheaves (Items 21 and 22)

Compressed air is generally measured in horse power (hp). A good rule of thumb is that 1 hp of compressed air is equal to 4 standard cubic feet per minute (SCFM). 1 standard cubic foot of air is measured after the air from the compressor expands back to atmospheric pressure. Therefore, if you use a $1/2$-hp motor then the best output you will be able to achieve is 2 SCFM. If you use a 1 $1/2$-hp motor then you will be able to achieve 6 SCFM.

Selecting the correct drive ratio is very important to best use the hp of the motor that you have obtained. Additionally, the wrong ratio may result in overloading the motor and severe damage may result. Determining the correct drive ratio for your compressor depends on three factors: (1) RPM of the motor, (2) the hp of the motor, and (3) the displacement of the engine/pump.

The following simple formulas may be used to determine the ratio that is required. Once the drive ratio is determined, a set of sheaves and vee belt may be selected.

1. Find the cubic foot displacement of the engine/pump.

$$[(Bore\ in. \div 2)^2 \times 3.14 \times stroke\ in.] \div 1728 = (CFD)$$

2. From the motor name plate, determine the RPM.

3. Calculate the SCFM with a motor/pump drive ratio of 1 to 1.

$$CFD \times RPM = SCFM$$

4. From the motor nameplate, determine the hp.

5. Calculate the ratio required.

$$SCFM \div (hp \times 4) = Drive\ Ratio$$

6. Calculate the appropriate set.

7. Motor sheave diameter × drive ratio = Diameter of engine/pump sheave

Example:

Bore	=	2.38 in.
Stroke	=	1.75 in.
Motor hp	=	0.75
Motor RPM	=	1725
Motor sheave diameter	=	2.00 in.

Note: 2.00-in. motor sheave diameter is an example; any diameter may be used.

1. $[(2.38 \div 2)^2 \times 3.14 \times 1.75] \div 1728 = 0.0045$ CFD

2. $0.0045 \times 1725 = 7.76$ SCFM

3. $7.76 \div (0.75 \times 4) = 2.59$ to 1 drive ratio

4. $2.00 \times 2.59 = 5.18$-in. diameter of engine/pump sheave

Assembly Instructions

1. Acquire everything listed in Table 15-1, "Air Compressor Bill of Materials."

2. Convert engine to pump per Table 15-2, "Engine Preparation Instructions."

3. Assemble the base per Fig. 15-2.

4. Drill item 46 per Fig. 15-4. (8 each)

5. Drill tank skirts as shown in Fig. 15-3.

6. Assemble the 8 mounting tabs (46) to the skirts of the tank per Fig. 15-3.

7. Attach the tank assembly to the base per Fig. 15-3.

8. Drill item 20 with a $7/16$-in. hole per Fig. 15-5.

9. Assemble items 2, 11, and 3 per Fig. 15-1.

10. Insert drilled item 20 into item 3 and tighten.

11. Slide item 55 into 20 until it bottoms in the tank, lift up $1/4$-in. and mark the location of 20.

12. Remove drilled item 20 and weld item 55 per Fig. 15-5.

13. After cooling, reinstall item 20/55 into 3 per Fig. 15-1.

14. Assemble items 4, 11, 12, and 15 per Fig. 15-1.

15. Assemble items 4, 6, 7, 8, and 16 per Fig. 15-1.

16. Drill 2-in. hole on center of the motor mount (39).

17. Bolt the motor mount to the tank/base assembly per Fig. 15-1.

18. Mount pump 42 to motor mount 39 per Fig. 15-1.

19. Install sheave 21 on the pump 42 and sheave 22 on the motor 43.

20. Place motor (43) on the motor mount (39) and position for the proper tension of the belt.

21. Mount the motor and tension the belt (23).

22. Connect the pump outlet 18 to fitting 16. (Use caution not to kink item 19 when bending).

23. Assemble item 9 into item 4 per Fig. 15-1.

24. Connect item 9 to item 17. (Use caution not to kink item 10 when bending).

25. Assemble items 5, 13, 14, 51, 52, and 53 per Fig. 15-1.

26. Mount three strain reliefs 33 into the electrical box 37 per Fig. 15-6.

27. Install the power cord 30 and 2 cables 32 into the electrical box.

28. Mount the electrical box 37 to the motor mount via item 50 (in Table 15-1, "Air compressor bill of materials") per Fig. 15-1.

29. Wire motor 43, pressure switch 9, and switch 38 per wiring diagram, Fig. 15-1, instructions on motor plate and instructions are provided with the pressure switch.

30. Adjust pressure switch per manufacturer's instructions to 60 psi maximum.

31. Turn on compressor.

32. After compressor turns off, check all fittings for leaks with soap bubbles.

33. Bleed down compressor and allow it to pump back up to verify pressure switch settings.

34. Place finished compressor in service.

Some Notes and Comments

There are two components that are recommended to be purchased new. The first is item 6, the pressure relief valve. This is a very important safety device and should be in perfect condition. The second is item 9, the pressure switch. It is unusual to find a used pressure switch and when one is found, they are normally in rather poor condition. Both of these components are inexpensive when purchased new and will provide you with a reliable and safe air compressor.

Take your time acquiring your components. Don't jump at every little part that may be usable. An old beat-up electric motor is no bargain if it's going to fail 3 months after you place the compressor in service. It is remarkable what you can obtain for little or no money if you are patient and diligent. Broaden your search to include: pawn shops, secondhand stores, junk yards, surplus centers, thrift stores, flea markets, church sales, garage sales, estate sales, your employer, your school, your family, your friends, your neighbors, and the like. Don't be afraid to say, "Can I have that?" (while pointing at a choice electric motor collecting dust in the corner of some gas station). More often than not the owner will say, "sure." Always be on the lookout for what you need and everything will come to you.

Once you have your compressor up and running, you'll want to get some air tools. First a $^3/_8$-in. ID air hose, about 50-ft long, with quick disconnects on both ends. An air chuck for filling tires and a blowgun are almost mandatory for any air compressor. If you do a lot of work on cars, a $^1/_2$-in. drive air wrench and a $^3/_8$-in. air ratchet are very handy tools. If you do a lot of body work or fabricating light metals, an air drill is quite useful. Appliance repair is made much easier with a $^3/_8$-in. drive air wrench. If you are a carpenter or a cabinetmaker, you will find great uses for air nailers and staplers. A siphon feed pressure paint gun does an excellent job of painting equipment and autos. For the paint gun you will need a venting regulator. These guns seem to work well with an air pressure of about 20 to 25 psi. If you want to paint cars it would be a good idea to equip the air line with desiccant filter/dryer or build the dryer set outlined in Chap. 16. When looking for air tools pay very close attention to the CFM requirements of the tool. Air tools that run continuously (like air sanders, grinders, and blasters) will require higher CFM and may overwork the compressor. A good rule of thumb is to use your compressor at no more than a 50 percent duty cycle. As an example, the compressor should not run more than a total of 5 minutes during any 10-minute period. Pushing the compressor for higher duty cycles may overheat the pump and cause damage. If you intend to use high CFM tools, then select a $1^1/_2$-hp motor when building the compressor.

Inspect your compressor and air tools regularly. Oil your air wrenches and drills by putting a few drops of oil into the air connection and run the tool for 5 seconds. Check the oil level in the pump. If the level gets low, top it off. Look at how clean the oil is and change as necessary. Check the belt tension. Periodically check the fittings to be sure that no leaks have developed. Drain the water out of the tank via item 5 ($^1/_4$-in. ball valve) every day of operation. If you take care of your compressed air system, it will take care of you.

16

Home-Built Compressed Air Drying System

Effectively drying compressed air usually requires rather expensive equipment. For small, home-shop compressors a dryer can be assembled with very little effort that will provide the same air quality that a typical refrigerated air dryer will produce.

The dryer is made by immersing a coil of $^3/_8$-in. copper tube, approximately 20 to 30 ft in length, into an ice bath as shown in Fig. 16-1. A simple trap assembly is attached to the output of the coil to control the separated water. Figure 16-2 shows how the trap assembly is constructed.

An ordinary styrofoam ice chest can be used for the bath. A piece of $^1/_2$-in. plywood is cut and drilled to support the coil assembly from the ice chest lid. A dryer like this will be able to carry approximately a 15-SCFM flow rate.

Figure 16-3 shows how the ice bath dryer is used with a compressor. These dryers provide an excellent solution for the home-shop compressor. They produce air quality that is appropriate for most compressed air applications. Typically, a 10-lb ice charge will last a full day if only one or two mechanics are using the system. As can be expected, the harder you push the compressor the quicker the ice will melt.

For applications that require a higher degree of drying, such as automobile paint and fine finishes, a second stage of drying is required. A dry ice dryer can be constructed that will easily provide 0°F to −40°F dew points. The principle of these dryers is that they freeze out any oil or water vapor that is being carried within the air stream. A central core is immersed in a dry ice slurry that is made up from crushed dry ice and acetone. As the compressed air passes through the cooled chamber, any water or oil that comes in contact with the walls is instantaneously

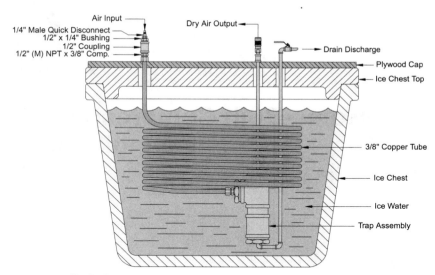

Air Input
1/4" Male Quick Disconnect
1/2" x 1/4" Bushing
1/2" Coupling
1/2" (M) NPT x 3/8" Comp.
Dry Air Output
Drain Discharge
Plywood Cap
Ice Chest Top
3/8" Copper Tube
Ice Chest
Ice Water
Trap Assembly

Figure 16-1   Ice bath dryer.

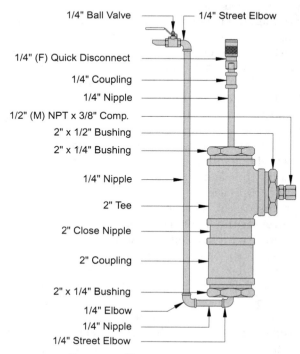

1/4" Ball Valve
1/4" Street Elbow
1/4" (F) Quick Disconnect
1/4" Coupling
1/4" Nipple
1/2" (M) NPT x 3/8" Comp.
2" x 1/2" Bushing
2" x 1/4" Bushing
1/4" Nipple
2" Tee
2" Close Nipple
2" Coupling
2" x 1/4" Bushing
1/4" Elbow
1/4" Nipple
1/4" Street Elbow

Figure 16-2   Trap assembly.

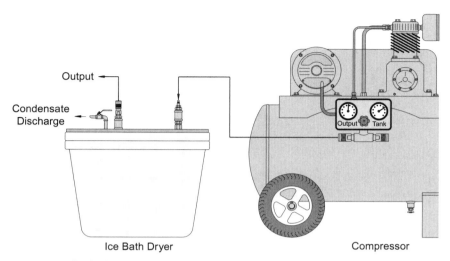

Output

Condensate
Discharge

Ice Bath Dryer

Compressor

Figure 16-3 Ice bath dryer/compressor setup.

frozen. As the air flows through the dryer, ice builds up on the inside walls of the 2-in. nipple. The dryer must be periodically thawed and drained or it will eventually become clogged with ice. Figure 16-4 can be used as a guide for constructing a dry ice dryer.

The dry ice dryer should be used as a second-stage dryer only. This is because it is what is referred to as a "getter pump." A getter pump captures, contaminates, and holds them or "gets" them. The dry ice freezes the water and oil vapor and holds that frozen material on the inside of the core. Because of this, the capacity of the dryer is limited. If the dryer is used with the direct output of the compressor, it will be grossly overloaded, rapidly fill, and clog up. This, in turn, will stop the flow of compressed air. Figure 16-5 shows how the dry ice dryer should be set up.

Caution: The temperature of dry ice is $-109°F$. Severe burns can occur if dry ice comes in contact with the skin. Always wear protective gloves and safety goggles when handling dry ice.

Caution: Acetone is extremely inflammable and should be handled with great care. Do not use acetone around any open flames. Always wear protective gloves and safety goggles when handling acetone. Always have access to proper fire control equipment when handling acetone.

Caution: Only use dry ice in adequately ventilated areas. Dry ice is frozen Co_2 and will produce Co_2 gas as it thaws. Co_2 does not support life and suffocation can occur if the dryer is not properly ventilated.

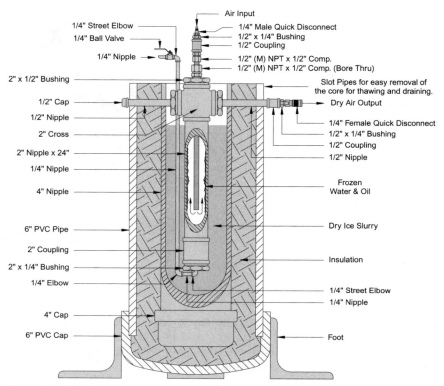

Figure 16-4 Dry ice dryer.

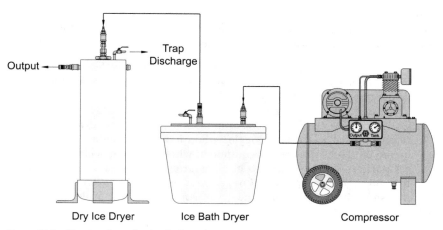

Figure 16-5 Dry ice dryer/ice bath dryer/compressor setup.

Glossary of Terms and Abbreviations

absorbed heat   Heat that is absorbed by the compression equipment.

aftercooler   The cooler that follows the output of the pump.

air over hydraulic   A hydraulic system that uses compressed air for power.

Approach temperature   The output temperature in reference to ambient.

ASCFM   Actual standard cubic feet per minute.

atm   Unit of pressure—the standard atmosphere—equal to 14.7 psi.

atmospheric pressure   14.7 psia.

bar   Metric pressure unit (1 bar = 14.5 psi).

blower   A compressor that is designed for high air movement at low pressures.

°C   Degrees Centigrade.

CFM   Cubic foot per minute (is not associated with any standard pressure).

check valve   A valve that allows flow in one direction only.

compress   The act of reducing the volume of a gas.

crow foot   Slang term for universal hose connector.

dew point   Temperature at which water vapor in air starts to condense.

drop trap   A trap intended to be installed at the bottom of a drop.

drop   A pipe that feeds air from the overhead line to floor level; a pipe that extends down from overhead piping, generally for an outlet.

dryer   A device for removing water vapor from a compressed air stream.

duplex   Dual.

expand   The act of increasing the volume of a gas.

°F Degrees Fahrenheit.

FNPT Female national pipe thread.

GH Garden hose.

hp Horse power (used to rate motors and air compression).

ID Inside diameter.

intercooler The cooler between stages of multistage pumps.

isolation valve A valve, intended to be used to isolate a portion of a system.

kW Kilowatt (1000 W) = 1.4 hp.

kWh Kilowatthour (1000 W for 1 h).

line trap A trap intended to be installed in a primary distribution pipeline.

MNPT Male national pipe thread.

motor controller A motor starter with all associated control circuitry.

motor starter A contactor used to start and stop an electric motor.

NPT National pipe thread.

OD Outside diameter.

pipe schedule (SCH) An industry standard for indicating pipe-wall thickness.

PLC Programmable logic controller.

pressure drop Pressure loss through piping or devices due to flow restrictions.

pressure switch A switch that activates in reference to pressure; a switch that is activated by pressure.

psi Pounds per square inch (atmospheric pressure reads 0 psi).

psia Pounds per square inch absolute (atmospheric pressure = 14.7 psi).

psig Pounds per square inch gauge (this is normally the same as psi).

pump The actual compressor, less any other equipment.

QD Quick disconnect.

receiver The pressure vessel that receives compressed air.

rejected heat Heat that is lost to the atmosphere.

SCBA Self-contained breathing apparatus.

SCFM Standard cubic foot per minute at atmospheric pressure.

SCUBA Self-contained underwater breathing apparatus.

short cycle The action of a motor turning on and off rapidly.

SNPT Straight national pipe threads.

two stage Pumps that compress air in two different steps.

unload The function of venting the pressure at the output of the pump.

unloader The valve that controls the unload function.

Miscellaneous Technical Information

V-Belt Tension

In order for your compressor to operate at peak efficiency, proper V-belt tension is essential. Additionally, improper tension, whether too tight or too loose, will accelerate the wear and reduce the service life of the belt. Tensioning the belt on any compressor should be carried out in accordance with the manufacturer's recommendation. In the absence of a recommendation, belts should be adjusted so that they have approximately $1/8$ in. to $1/4$ in. of deflection when pushed with one finger. For larger belts and multiple belts, a jack may be placed between the sheaves to achieve proper tension before tightening the motor bolts. Figure B-1 may be used as a guide for V-belt tensioning.

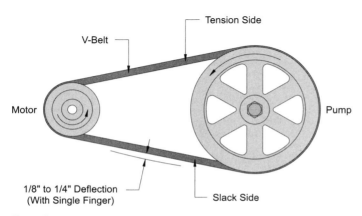

Figure B-1

Multiple Pipe Size Equivalents

In cases where double pipes are operating side by side, it is useful to be able to approximate the equivalent pipe size. Table B-1 shows the equivalent pipe size of two smaller pipes operating in parallel.

TABLE B-1

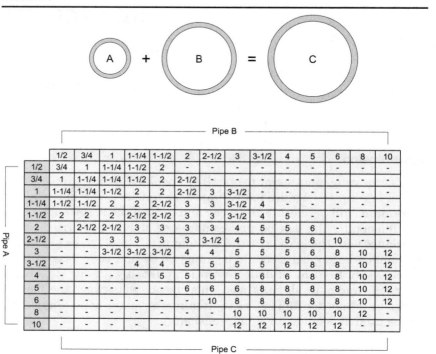

Pipe B

Pipe A \ Pipe B	1/2	3/4	1	1-1/4	1-1/2	2	2-1/2	3	3-1/2	4	5	6	8	10
1/2	3/4	1	1-1/4	1-1/2	2	-	-	-	-	-	-	-	-	-
3/4	1	1-1/4	1-1/4	1-1/2	2	2-1/2	-	-	-	-	-	-	-	-
1	1-1/4	1-1/4	1-1/2	2	2	2-1/2	3	3-1/2	-	-	-	-	-	-
1-1/4	1-1/2	1-1/2	2	2	2-1/2	3	3	3-1/2	4	-	-	-	-	-
1-1/2	2	2	2	2-1/2	2-1/2	3	3	3-1/2	4	5	-	-	-	-
2	-	2-1/2	2-1/2	3	3	3	3	4	5	5	6	-	-	-
2-1/2	-	-	3	3	3	3	3-1/2	4	5	5	6	10	-	-
3	-	-	3-1/2	3-1/2	3-1/2	4	4	5	5	5	6	8	10	12
3-1/2	-	-	-	4	4	5	5	5	5	6	8	8	10	12
4	-	-	-	-	5	5	5	5	6	6	8	8	10	12
5	-	-	-	-	-	6	6	6	8	8	8	8	10	12
6	-	-	-	-	-	-	10	8	8	8	8	8	10	12
8	-	-	-	-	-	-	-	10	10	10	10	10	12	-
10	-	-	-	-	-	-	-	12	12	12	12	12	-	-

Pipe C

Pressure/Temperature Ratings for Commercial Pipe Fittings

The pressure ratings on 150- and 300-psi-class fittings can vary with temperature. The nominal rating of the fitting is at an elevated temperature, and these fittings can be used for higher pressures if the temperature is carefully controlled. Class fittings of 150 psi have a working temperature of 350°F and class fittings of 300 psi have a working temperature of 550°F. If these fittings are only exposed to lower temperatures, then their working pressure can be considerably higher than their nominal rating. Table B-2 shows acceptable pressures at various temperatures for 150- and 300-psi-class malleable iron fittings. Figure B-2 shows the temperature conversion chart and Fig. B-3 shows pressure calculation.

TABLE B-2

Temperature		Class			
		150	300		
°F	°C		1/4" to 1"	1-1/4" to 2"	2-1/2-3"
-20° to 150°	-29° to 66°	300	2000	1500	1000
200°	93°	265	1785	1350	910
250°	121°	225	1575	1200	825
300°	149°	185	1360	1050	735
350°	177°	150	1150	900	650
400°	204°	-	935	750	560
450°	232°	-	725	600	475
500°	260°	-	510	450	385
550°	288°	-	300	300	300
		Working Pressure in psi			

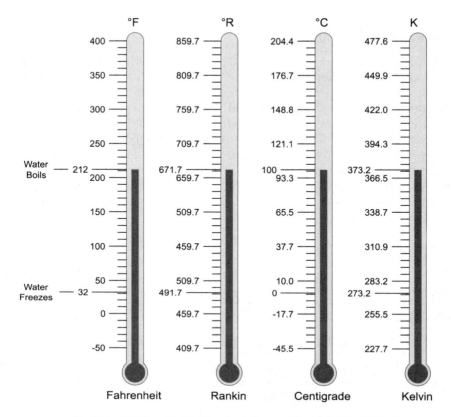

To Convert Fahrenheit to Centigrade: °C = [5 X (°F - 32)] ÷ 9
To Convert Centigrade to Fahrenheit: °F = [9 x (°C ÷ 5)] + 32
To Convert Fahrenheit to Rankine: °R = F + 459.7
To Convert Centigrade to Kelvin: K = °C + 273.2

Figure B-2

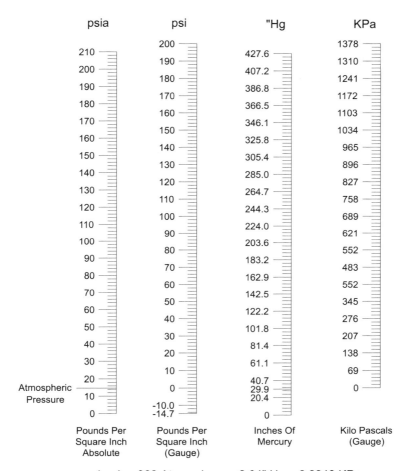

1 psi = .068 Atmosphere = 2.04" Hg = 6.8948 KPa

Figure B-3

Sheave Location and Alignment

When servicing the sheaves and belt(s) on the pump and motor, it is important to have them properly installed. Proper installation will ensure that the bearings and drive belt provide the service life that they are intended to. Figure B-4 shows proper alignment of the motor sheave. The preferred alignment will place the vee as close as possible to the motor bearing. This will place the lowest possible load on the motor bearing and consequently provide a longer service life. Acceptable locations are in the middle of the shaft while unacceptable locations are any positions where the shaft does not extend at least $1/4$ in. out of the sheave hub.

Alignment of the belt is just as important as tensioning. Misaligned belts will not transmit the required horsepower and will wear out prematurely. To align the belt, use a straightedge as shown in Fig. B-5. Minor alignment may be accomplished by adjusting the position of the motor sheave. To correct for major misalignment, the motor position must be adjusted.

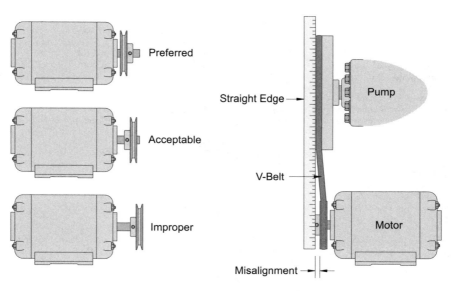

Figure B-4   **Figure B-5**

Taper Pipe Thread Engagement

When cutting and threading pipe, it is important to know the engagement length of the threads. Figure B-6 can be used as a guide for the thread engagement of NPT joints. Be aware that hand-tight engagements can be affected by threads that are dirty, damaged, or have slight irregularities.

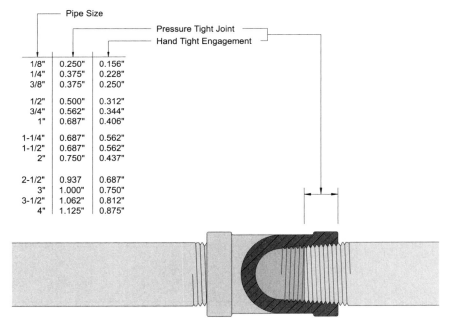

Figure B-6

TABLE B-3

Cross

	1/8	1/4	3/8	1/2	3/4	1	1-1/4	1-1/2	2	2-1/2	3	3-1/2	4	5	6
A	0.69	0.81	0.94	1.12	1.31	1.50	1.75	1.94	2.25	2.69	3.06	-	3.81	-	-

Tee

	1/8	1/4	3/8	1/2	3/4	1	1-1/4	1-1/2	2	2-1/2	3	3-1/2	4	5	6
A	0.69	0.81	0.94	1.12	1.31	1.50	1.75	1.94	2.25	2.69	3.06	3.56	3.81	4.50	5.12

Lateral Tee

	1/8	1/4	3/8	1/2	3/4	1	1-1/4	1-1/2	2	2-1/2	3	3-1/2	4	5	6
T	-	-	0.50	0.62	0.75	0.88	1	1.12	1.25	1.50	1.69	-	2	-	-
U	-	-	1.56	1.69	2.06	2.56	2.94	3.25	3.94	4.75	5.56	-	7	-	-
V	-	-	1.94	2.31	2.81	3.31	3.94	4.75	5.19	6.25	7.25	-	9	-	-

90° Elbow

	1/8	1/4	3/8	1/2	3/4	1	1-1/4	1-1/2	2	2-1/2	3	3-1/2	4	5	6
A	0.69	0.81	0.94	1.12	1.31	1.50	1.75	1.94	2.25	2.69	3.06	3.56	3.81	4.50	5.12

90° Street Elbow

	1/8	1/4	3/8	1/2	3/4	1	1-1/4	1-1/2	2	2-1/2	3	3-1/2	4	5	6
A	0.69	0.81	0.94	1.12	1.31	1.50	1.75	1.94	2.25	2.69	3.06	3.56	-	-	-
J	1	1.19	1.56	1.62	1.88	2.12	2.56	2.69	3.25	3.88	4.25	5.69	-	-	-

45° Elbow

	1/8	1/4	3/8	1/2	3/4	1	1-1/4	1-1/2	2	2-1/2	3	3-1/2	4	5	6
C	0.69	0.75	0.81	0.88	1	1.12	1.31	1.56	1.69	1.94	2.19	-	2.62	3.06	3.56

45° Street Elbow

	1/8	1/4	3/8	1/2	3/4	1	1-1/4	1-1/2	2	2-1/2	3	3-1/2	4	5	6
C	0.69	0.75	0.81	0.88	1	1.12	1.31	1.56	1.69	-	-	-	-	-	-
K	0.88	0.94	1	1.12	1.31	1.56	1.69	1.88	2.25	-	-	-	-	-	-

Coupling

	1/8	1/4	3/8	1/2	3/4	1	1-1/4	1-1/2	2	2-1/2	3	3-1/2	4	5	6
W	0.94	1.06	1.19	1.31	1.50	1.69	1.94	2.12	2.50	2.88	3.19	-	3.69	-	-

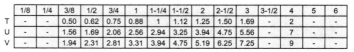

(Continued)

TABLE B-3 (*Continued*)

Union

1/8	1/4	3/8	1/2	3/4	1	1-1/4	1-1/2	2	2-1/2	3	3-1/2	4	5	6
R														
1.31	1.81	1.81	1.94	2.06	2.06	2.62	2.75	2.94	3.62	3.25	-	-	-	-

Cap

1/8	1/4	3/8	1/2	3/4	1	1-1/4	1-1/2	2	2-1/2	3	3-1/2	4	5	6

Return Bend

1/8	1/4	3/8	1/2	3/4	1	1-1/4	1-1/2	2	2-1/2	3	3-1/2	4	5	6
L														
0.69	0.75	0.81	0.88	1	1.12	1.31	1.56	1.69	1.94	2.19	-	2.62	3.06	3.56

Plugs & Bushings

1/8	1/4	3/8	1/2	3/4	1	1-1/4	1-1/2	2	2-1/2	3	3-1/2	4	5	6

TABLE B-4

Pipe Size	Inside Diameter	Area of Bore
1/2	0.622	0.304
3/4	0.824	0.533
1	1.049	0.864
1-1/4	1.380	1.496
1-1/2	1.610	2.036
2	2.067	3.356
2-1/2	2.469	4.788
3	3.068	7.393
3-1/2	3.548	9.887
4	4.026	12.730
5	5.047	20.000
6	6.065	28.890
8	7.981	50.027
10	10.020	78.854
12	12.000	113.097

TABLE B-5 Sizes, Schedules, Wall Thickness, Outside and Inside Diameters for 1/8- through 12-in. Commercial Pipe Per ASA-B36.10 and B36.19

Pipe size	Outside dia.	Wall thickness and inside diameter							
		std./40	ID	XS/80	ID	160	ID	XX	ID
1/8	0.405	0.068	0.269	0.095	0.215	—	—	—	—
1/4	0.540	0.088	0.364	0.119	0.302	—	—	—	—
3/8	0.675	0.091	0.493	0.126	0.423	—	—	—	—
1/2	0.840	0.109	0.622	0.147	0.546	0.188	0.464	0.294	0.252
3/4	1.050	0.113	0.824	0.154	0.742	0.219	0.612	0.308	0.434
1	1.315	0.133	1.049	0.179	0.957	0.250	0.815	0.358	0.599
1-1/4	1.660	0.140	1.380	0.191	1.278	0.250	1.160	0.382	0.896
1-1/2	1.900	0.145	1.610	0.200	1.500	0.281	1.338	0.400	1.100
2	2.375	0.154	2.067	0.218	1.939	0.344	1.687	0.436	1.503
2-1/2	2.875	0.203	2.469	0.276	2.323	0.375	2.125	0.552	1.771
3	3.500	0.216	3.068	0.300	2.900	0.437	2.625	0.600	2.300
3-1/2	4.000	0.226	3.548	0.318	3.364	—	—	—	—
4	4.500	0.237	4.026	0.337	3.826	0.531	3.438	0.674	3.152
5	5.563	0.258	5.047	0.375	4.813	0.625	4.313	0.750	4.063
6	6.625	0.280	6.065	0.432	5.761	0.719	5.187	0.864	4.897
8	8.625	0.322	7.981	0.500	7.625	0.906	6.813	0.875	6.875
10	10.750	0.365	10.020	0.500	9.750	1.125	8.500	1.000	8.750
				0.594	9.562				
12	12.750	0.375	12.000	0.500	11.750	1.312	10.125	1.000	10.750
		0.406	11.938	0.688	11.375				

TABLE B-6 Weight per Foot for 1/8- through 12-in. Commercial Pipe

Weight per Foot (lb)

Pipe size	Std./40	W/water	XS/80	W/water	160	W/water	XX	W/water
1/8	0.244	+0.025	.314	+0.016	–	–	–	–
1/4	0.424	+0.045	.535	+0.031	–	–	–	–
3/8	0.567	+0.083	.738	+0.061	–	–	–	–
1/2	0.850	+0.132	1.087	+0.101	1.310	+0.074	1.714	+0.022
3/4	1.130	+0.230	1.473	+0.186	1.940	+0.128	2.440	+0.063
1	1.678	+0.374	2.171	+0.311	2.850	+0.226	3.659	+0.122
1-1/4	2.272	+0.647	2.996	+0.555	3.764	+0.458	5.214	+0.273
1-1/2	2.717	+0.882	3.631	+0.765	4.862	+0.608	6.408	+0.412
2	3.652	+1.452	5.022	+1.279	7.450	+0.970	9.026	+0.769
2-1/2	5.790	+2.072	7.660	+1.834	10.010	+1.535	13.690	+1.067
3	7.570	+3.200	10.250	+2.860	14.320	+2.340	18.580	+1.800
3-1/2	9.110	+4.280	12.510	+3.850	–	–	22.850	+2.530
4	10.790	+5.510	14.980	+4.980	22.600	+4.020	27.540	+3.380
5	14.620	+8.660	20.780	+7.870	32.960	+6.320	38.550	+5.620
6	18.970	+12.510	28.570	+11.290	45.300	+9.160	53.160	+8.140
8	28.550	+21.600	43.390	+19.800	74.700	+15.700	72.420	+16.100
10	40.480	+34.100	64.400	+31.100	116.00	+24.600	–	–
12	–	–	88.600	+44.000	161.00	+34.900	125.50	+39.300
12" Std.:	49.600	+48.900						
12 Schedule 40:	53.600	+48.500						

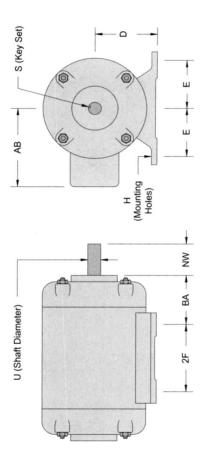

U (Shaft Diameter)

S (Key Set)

AB

D

E

E

H
(Mounting
Holes)

NW

BA

2F

TABLE B-7

Frame	D	E	2F	H	U	BA	NW	S
48	3.00	2.12	2.75	0.30	0.50	2.50	1.50	Flat
56	3.50	2.44	3.00	0.30	0.63	2.80	1.88	0.19
143	3.50	2.75	4.00	0.30	0.75	2.30	2.00	0.19
143T	3.50	2.75	4.00	0.30	0.88	2.30	2.25	0.19
145	3.50	2.75	5.00	0.30	0.75	2.30	2.00	0.19
145T	3.50	2.75	5.00	0.30	0.88	2.30	2.25	0.19
182	4.50	3.75	4.50	0.40	0.88	2.80	2.25	0.19
182T	4.50	3.75	4.50	0.40	1.13	2.80	2.75	0.25
184	4.50	3.75	5.50	0.40	0.88	2.80	2.25	0.19
184T	4.50	3.75	5.50	0.40	1.13	2.80	2.75	0.25
203	5.00	4.00	5.50	0.40	0.75	3.10	2.25	0.19
204	5.00	4.00	6.50	0.40	0.75	3.10	2.25	0.19
213	5.30	4.25	5.50	0.40	1.13	3.50	3.00	0.25
213T	5.30	4.25	5.50	0.40	1.38	3.50	3.38	0.31
215	5.30	4.25	7.00	0.40	1.13	3.50	3.00	0.25
215T	5.30	4.25	7.00	0.40	1.38	3.50	3.38	0.31
224	5.50	4.50	6.75	0.40	1.00	3.50	3.00	0.25
225	5.50	4.50	7.50	0.40	1.00	3.50	3.00	0.25
254	6.30	5.00	8.25	0.50	1.13	4.30	3.37	0.25
254U	6.30	5.00	8.25	0.50	1.38	4.30	3.75	0.31
254T	6.30	5.00	8.25	0.50	1.63	4.30	4.00	0.38
256U	6.30	5.00	10.00	0.50	1.38	4.30	3.75	0.31
256T	6.30	5.00	10.00	0.50	1.63	4.30	4.00	0.38
284	7.00	5.50	9.50	0.50	1.25	4.80	3.75	0.25
284U	7.00	5.50	9.50	0.50	1.63	4.80	4.88	0.38
284T	7.00	5.50	9.50	0.50	1.88	4.80	4.62	0.50
284TS	7.00	5.50	9.50	0.50	1.63	4.80	3.25	0.38
286U	7.00	5.50	11.00	0.50	1.63	4.80	4.88	0.38
286T	7.00	5.50	11.00	0.50	1.88	4.80	4.62	0.50

TABLE B-7 *(Continued)*

Frame	D	E	2F	H	U	BA	NW	S
286TS	7.00	5.50	11.00	0.50	1.63	4.80	3.25	3.75
324	8.00	6.25	10.50	0.70	1.63	5.30	4.87	0.38
324U	8.00	6.25	10.50	0.70	1.88	5.30	5.62	0.50
324S	8.00	6.25	10.50	0.70	1.63	5.30	3.25	0.38
324T	8.00	6.25	10.50	0.70	2.13	5.30	5.25	0.50
324TS	8.00	6.25	10.50	0.70	1.88	5.30	3.75	0.50
326	8.00	6.25	12.00	0.70	1.63	5.30	4.87	0.38
326U	8.00	6.25	12.00	0.70	1.88	5.30	5.62	0.50
326S	8.00	6.25	12.00	0.70	1.63	5.30	3.25	0.38
326T	8.00	6.25	12.00	0.70	2.13	5.30	5.25	0.50
326TS	8.00	6.25	12.00	0.70	1.88	5.30	3.75	0.50
364	9.00	7.00	11.30	0.70	1.88	5.90	5.62	0.50
364S	9.00	7.00	11.30	0.70	1.63	5.90	3.25	0.38
364U	9.00	7.00	11.30	0.70	2.13	5.90	6.37	0.50
364US	9.00	7.00	11.30	0.70	1.88	5.90	3.75	0.50
364T	9.00	7.00	11.30	0.70	2.38	5.90	5.88	0.63
364TS	9.00	7.00	11.30	0.70	1.88	5.90	3.75	0.50
365	9.00	7.00	12.30	0.70	1.88	5.90	5.62	0.50
365S	9.00	7.00	12.30	0.70	1.63	5.90	3.25	0.38
365U	9.00	7.00	12.30	0.70	2.13	5.90	6.37	0.50
365US	9.00	7.00	12.30	0.70	1.88	5.90	3.75	0.50
365T	9.00	7.00	12.30	0.70	2.38	5.90	5.88	0.63
365TS	9.00	7.00	12.30	0.70	1.88	5.90	3.75	0.50
404	10.00	8.00	12.30	0.80	2.13	6.60	6.37	0.50
404S	10.00	8.00	12.30	0.80	1.88	6.60	3.75	0.50
404U	10.00	8.00	12.30	0.80	2.38	6.60	7.12	0.63
404US	10.00	8.00	12.30	0.80	2.13	6.60	4.25	0.50
404T	10.00	8.00	12.30	0.80	2.88	6.60	7.25	0.75
404TS	10.00	8.00	12.30	0.80	2.13	6.60	4.25	0.50

(Continued)

TABLE B-7 *(Continued)*

Frame	D	E	2F	H	U	BA	NW	S
405	10.00	8.00	13.80	0.80	2.13	6.60	6.37	0.50
324	8.00	6.25	10.50	0.70	1.63	5.30	4.87	0.38
324U	8.00	6.25	10.50	0.70	1.88	5.30	5.62	0.50
405S	10.00	8.00	13.80	0.80	1.88	6.60	3.75	0.50
405U	10.00	8.00	13.80	0.80	2.38	6.60	7.12	0.63
405US	10.00	8.00	13.80	0.80	2.13	6.60	4.25	0.50
405T	10.00	8.00	13.80	0.80	2.88	6.60	7.25	0.75
405TS	10.00	8.00	13.80	0.80	2.13	6.60	4.25	0.50
444	11.00	9.00	14.50	0.80	2.38	7.50	7.12	0.63
444S	11.00	9.00	14.50	0.80	2.13	7.50	4.25	0.50
444U	11.00	9.00	14.50	0.80	2.88	7.50	8.62	0.75
444US	11.00	9.00	14.50	0.80	2.13	7.50	4.25	0.50
444T	11.00	9.00	14.50	0.80	3.38	7.50	8.50	0.88
444TS	11.00	9.00	14.50	0.80	2.38	7.50	4.75	0.63
445	11.00	9.00	16.50	0.80	2.38	7.50	7.12	0.63
445S	11.00	9.00	16.50	0.80	2.13	7.50	4.25	0.50
445U	11.00	9.00	16.50	0.80	2.88	7.50	8.62	0.75
445US	11.00	9.00	16.50	0.80	2.13	7.50	4.25	0.50
445T	11.00	9.00	16.50	0.80	3.38	7.50	8.50	0.88
445TS	11.00	9.00	16.50	0.80	2.38	7.50	4.75	0.63
447TS	11.00	9.00	20.00	*	*	*	*	*
449TS	11.00	9.00	25.00	*	*	*	*	*
504U	13.00	10.00	16.00	0.90	2.88	8.50	8.62	0.75
505	13.00	10.00	18.00	0.90	2.88	8.50	8.62	0.75
505S	13.00	10.00	18.00	0.90	2.13	8.50	4.25	0.50

*Per manufacturer's specification

TABLE B-8   Tap Drill Sizes for National Pipe Threads (NPT)

Pipe size	Threads per inch	Drill size	Drill size w/ taper pipe reamer	Drill size straight pipe
1/16	27	D	15/64	1/4
1/8	27	Q	21/64	11/32
1/4	18	7/16	27/64	29/64
3/8	18	37/64	9/16	37/64
1/2	14	23/32	11/16	23/32
3/4	14	59/64	57/64	15/16
1	11-1/2	1-5/32	1-1/8	1-11/64
1-1/4	11-1/2	1-1/2	1-15/32	1-33/64
1-1/2	11-1/2	1-47/64	1-23/32	1-3/4
2	11-1/2	–	2 3/16	–

TABLE B-9   Tap Drill Sizes and Basic Thread Information

Tap size	UNF/ UNC	Threads per inch	Basic major diameter (inches)	Basic minor diameter (inches)	Drill size
0-80	UNF	80	0.0600	0.0447	3/64
1-64	UNC	64	0.0730	0.0538	#54
2-56	UNC	56	0.0860	0.0641	#50
2-64	UNF	64	0.0860	0.0668	#50
4-40	UNC	40	0.1120	0.0813	#43
4-48	UNF	48	0.1120	0.0864	#42
5-40	UNC	40	0.1250	0.0943	#38
5-44	UNF	44	0.1250	0.0971	#37
6-32	UNC	32	0.1360	0.0997	#36
6-40	UNF	40	0.1360	0.1073	#33
8-32	UNC	32	0.1640	0.1257	#29
8-36	UNF	36	0.1640	0.1299	#29
10-24	UNC	24	0.1900	0.1389	#25
10-32	UNF	32	0.1900	0.1517	#21
1/4-20	UNC	20	0.2500	0.1887	#7
1/4-28	UNF	28	0.2500	0.2062	#3
5/16-18	UNC	18	0.3125	0.2443	F
5/16-24	UNF	24	0.3125	0.2614	I
3/8-16	UNC	16	0.3750	0.2983	5/16
3/8-24	UNF	24	0.3750	0.3239	Q
7/16-14	UNC	14	0.4375	0.3499	U
7/16-20	UNF	20	0.4375	0.3762	25/64
1/2-13	UNC	13	0.5000	0.4056	27/64
1/2-20	UNF	20	0.5000	0.4387	29/64
9/16-12	UNC	12	0.5625	0.4603	31/64
9/16-18	UNF	18	0.5625	0.4943	33/64
5/8-11	UNC	11	0.6250	0.5135	17/32
5/8-18	UNF	18	0.6250	0.5568	37/64
3/4-10	UNC	10	0.7500	0.6273	21/32
3/4-16	UNF	16	0.7500	0.6733	11/16
7/8-9	UNC	9	0.8750	0.7387	49/64
7/8-14	UNF	14	0.8750	0.7874	13/16
1-8	UNC	8	1.000	0.8466	7/8
1-14	UNF	14	1.000	0.8978	15/16
1 1/18-7	UNC	7	1.1250	0.9497	63/64
1 1/8-12	UNF	12	1.1250	1.0228	1 3/64
1 1/4-7	UNC	7	1.2500	1.0747	1 7/64
1 3/8-6	UNC	6	1.3750	1.1705	1 13/64
1 1/2-6	UNC	6	1.5000	1.2955	1 11/32
1 3/4-5	UNC	5	1.7500	1.5046	1 35/64
2-4-1/2	UNC	4 1/2	2.0000	1.7274	1 25/32

TABLE B-10   **Decimal Equivalents**

English	Metric	Decimal	English	Metric	Decimal
–	0.1	0.0039	59	–	0.0410
–	0.2	0.0079	58	–	0.0420
–	0.3	0.0118	57	–	0.0430
80	–	0.0135	56	–	0.0465
79	–	0.0145	3/64	–	0.0469
1/64	–	0.0156	55	–	0.0520
–	0.4	0.0157	54	–	0.0550
78	–	0.0160	53	–	0.0595
77	–	0.0180	1/16	–	0.0625
–	0.5	0.0197	52	–	0.0635
76	–	0.0200	51	–	0.0670
75	–	0.0210	50	–	0.0700
74		0.0225	49	–	0.0730
–	0.6	0.0236	48	–	0.0760
73	–	0.0240	5/64	–	0.0781
72	–	0.0250	47	–	0.0785
71	–	0.0260	–	2	0.0787
–	0.7	0.0276	46	–	0.0810
70	–	0.0280	45	–	0.0820
69	–	0.0292	44	–	0.0860
68	–	0.0310	43	–	0.0890
1/32	–	0.0312	42	–	0.0935
–	0.8	0.0315	3/32	–	0.0937
67	–	0.0320	41	–	0.0960
66	–	0.0330	40	–	0.0980
65	–	0.0350	39	–	0.0995
–	0.9	0.0354	38	–	0.1015
64	–	0.0360	37	–	0.1040
63	–	0.0370	36	–	0.1065
62	–	0.0380	7/64	–	0.1094
61	–	0.0390	35	–	0.1100
–	1	0.0394	34	–	0.1110
60	–	0.0400	33	–	0.1130
32	–	0.1160	6	–	0.2040
–	3	0.1181	5	–	0.2055
31	–	0.1200	4	–	0.2090
1/8	–	0.1250	3	–	0.2130
30	–	0.1285	7/32	–	0.2187
29	–	0.1360	2	–	0.2210
28	–	0.1405	1	–	0.2280
9/64	–	0.1406	A	–	0.2340
27	–	0.1440	15/64	–	0.2344
26	–	0.1470	–	6	0.2362
25	–	0.1495	B	–	0.2380
24	–	0.1520	C	–	0.2420
23	–	0.1540	D	–	0.2460
5/32	–	0.1562	1/4	–	0.2500
22	–	0.1570	F	–	0.2570
–	4	0.1575	G	–	0.2610
21	–	0.1590	17/64	–	0.2656
20	–	0.1610	H	–	0.2660

(Continued)

TABLE B-10 Decimal Equivalents (*Continued*)

English	Metric	Decimal	English	Metric	Decimal
19	–	0.1660	I	–	0.2720
18	–	0.1695	–	7	0.2756
11/64	–	0.1719	J	–	0.2770
17	–	0.1730	K	–	0.2810
16	–	0.1770	9/32	–	0.2812
15	–	0.1800	L	–	0.2900
14	–	0.1820	M	–	0.2950
13	–	0.1850	19/64	–	0.2969
3/16	–	0.1875	N	–	0.3020
12	–	0.1890	5/16	–	0.3125
11	–	0.1910	–	8	0.3150
10	–	0.1935	O	–	0.3160
9	–	0.1960	P	–	0.3230
–	5	0.1968	21/64	–	0.3281
8	–	0.1990	Q	–	0.3320
7	–	0.2010	R	–	0.3390
13/64	–	0.2031	11/32	–	0.3437
S	–	0.3480	41/64	–	0.6406
–	9	0.3543	21/32	–	0.6562
T	–	0.3580	–	17	0.6693
23/64	–	0.3594	43/64	–	0.6719
U	–	0.3680	11/16	–	0.6875
3/8	–	0.3750	45/64	–	0.7031
V	–	0.3770	–	18	0.7087
W	–	0.3860	23/32	–	0.7187
25/64	–	0.3906	47/64	–	0.7344
–	10	0.3937	–	19	0.7480
X	–	0.3970	3/4	–	0.7500
Y	–	0.4040	49/64	–	0.7656
13/32	–	0.4062	25/32	–	0.7812
Z	–	0.4130	–	20	0.7874
27/64	–	0.4219	51/64	–	0.7969
–	11	0.4331	13/16	–	0.8125
7/16	–	0.4375	–	21	0.8268
29/64	–	0.4531	53/64	–	0.8281
15/32	–	0.4687	27/32	–	0.8437
–	12	0.4724	55/64	–	0.8594
31/64	–	0.4844	–	22	0.8661
1/2	–	0.5000	7/8	–	0.8750
–	13	0.5118	57/64	–	0.8906
33/64	–	0.5156	–	23	0.9055
17/32	–	0.5312	29/32	–	0.9062
35/64	–	0.5469	59/64	–	0.9219
–	14	0.5512	15/16	–	0.9375
9/16	–	0.5625	–	24	0.9449
37/64	–	0.5781	61/64	–	0.9531
–	15	0.5906	31/32	–	0.9687
19/32	–	0.5937	–	25	0.9842
39/64	–	0.6094	63/64	–	0.9844
5/8	–	0.6250	1	25.4	1.0000
–	16	0.6299			

TABLE B-11 Common Conversion Multipliers

Convert	Into	Multiply by
Absolute temp. °C–273	Temperature (°C)	1.0
Absolute temp. °F–460	Temperature (°F)	1.0
Atmospheres	Inches of mercury	29.92
Atmospheres	Kilograms/square centimeters	1.033
Atmospheres	Pounds/square inch (psi)	14.62
Bars.	Pounds/square inch	14.50
British thermal unit/minute	Horsepower	0.02356
British thermal unit/minute	Kilowatts	0.01757
British thermal unit/minute	Watts	17.57

C

Centimeters	Feet	3.281×10 E-2
Centimeters	Inches	0.3937
Centimeters	Kilometers	10 E-5
Centimeters	Meters	0.01
Centimeters	Miles	6.214×10 E-6
Centimeters	Millimeters	10.0
Centimeters	Yards	1.094×10 E-2
Cubic centimeters	Cubic feet	3.531×10 E-4
Cubic centimeters	Cubic inches	0.06102
Cubic centimeters	Cubic meters	10 E-6
Cubic centimeters	Cubic yards	1.308×10 E-6
Cubic centimeters	Gallons (U.S. liquid)	2.642×10 E-4
Cubic centimeters	Liters	0.001
Cubic centimeters	Pints (U.S. liquid)	2.113×10 E-3
Cubic centimeters	Quarts (U.S. liquid)	1.057×10 E-3
Cubic feet	Cubic centimeters	28,320.0
Cubic feet	Cubic inches	1,728.0
Cubic feet	Cubic meters	0.02832
Cubic feet	Cubic yards	0.03704
Cubic feet	Gallons (U.S. liquid)	7.48052
Cubic feet	Liters	28.32
Cubic feet	Pints (U.S. liquid)	59.84
Cubic feet	Quarts (U.S. liquid)	29.92
Cubic feet/second	Liters/minute	1699
Cubic inches	Cubic centimeters	16.39
Cubic inches	Cubic feet	5.787×10 E-4
Cubic inches	Cubic meters	1.639×10 E-5
Cubic inches	Cubic yards	2.143×10 E-5
Cubic inches	Gallons (U.S. liquid)	4.329×10 E-3
Cubic inches	Liters	0.01639
Cubic inches	Pints (U.S. liquid)	0.03463
Cubic inches	Quarts (U.S. liquid)	0.01732
Cubic meters	Cubic centimeters	10 E6
Cubic meters	Cubic feet	35.31

(Continued)

TABLE B-11   Common Conversion Multipliers (*Continued*)

Convert	Into	Multiply by
Cubic meters	Cubic inches	61,023.0
Cubic meters	Cubic yards	1.308
Cubic meters	Gallons (U.S. liquid)	264.2
Cubic meters	Liters	1,000.0
Cubic yards	Cubic centimeters	2832
Cubic yards	Cubic feet	27
Cubic yards	Cubic inches	46656
Cubic yards	Cubic meters	.768
Cubic yards	Gallons (gal)	201.98
Cubic yards	Liters	764.53
Cubic yards	Quarts	807.75

F

Feet	Centimeters	30.48
Feet	Kilometers	3.048×10 E-4
Feet	Meters	0.3048
Feet	Millimeters	304.8
Feet/minute	Meters/minute	0.3048
Feet/minute	Meters/second	0.00508
Feet of water	Atmospheres	0.02950
Feet of water	Inches of mercury	0.8826
Feet of water	Pounds/square inch	0.4335
Foot-pounds/minute	Horsepower	0.0000303
Foot-pounds/second	Kilowatts	0.00136
Foot-pounds	Kilogram-meters	0.1383

G

Gallons	Cubic yards	4.951×10 E-3
Gallons	Liters	3.785
Gallons of water	Pounds of water	8.3453
Gallons/second	Liters/minute	227.12
Grams	Kilograms	0.001
Grams	Ounces	0.0356
Grams	Pounds	0.0022

H

Horsepower	Foot-pounds/minute	33,000
Horsepower	Kilowatts	0.7457
Horsepower	Watts	745.7

I

Inches	Centimeters	2.540
Inches	Meters	2.540×10 E-2
Inches	Miles	1.578×10 E-5
Inches	Millimeters	25.40
Inches of mercury	Atmospheres	0.03342
Inches of mercury	Feet of water	1.133

TABLE B-11   Common Conversion Multipliers (*Continued*)

Convert	Into	Multiply by
Inches of mercury	Kilograms/square centimeter	0.03453
Inches of mercury	Pounds/square inch	0.4912
K		
Kilograms	Grams	1,000.0
Kilograms	Pounds	2.205
Kilograms/square centimeter	Atmospheres	0.9678
Kilograms/square centimeter	Inches of mercury	28.96
Kilograms/square centimeter	Pounds/square inch	14.22
Kilograms/square meter	Atmospheres	9.678×10 E-5
Kilograms/square meter	Bars	98.07×10 E-6
Kilograms/square meter	Inches of mercury	2.896×10 E-3
Kilogram-meters	Foot-pounds	7.233
Kilometers	Feet	3,281
Kilometers	Inches	3.937×10 E4
Kilometers	Miles	0.6214
Kilometers/hours	Miles/hour	0.6214
Kilowatts	British thermal unit/minute	56.92
Kilowatts	Foot-pounds/second	737.6
Kilowatts	Horsepower	1.341
Kilometers/hour		
Knots		
L		
Liters	Cubic centimeter	1,000.0
Liters	Cubic feet	0.03531
Liters	Cubic inches	61.02
Liters	Cubic meters	0.001
Liters	Cubic yards	1.308×10 E-3
Liters	Gallons (U.S. liquid)	0.2642
Liters	Pints (U.S. liquid)	2.113
Liters	Quarts (U.S. liquid)	1.057
Liters/minute	Cubic feet/second	5.886×10 E-4
Liters/minute	Gallons/seconds	4.403×10 E-3
M		
Meters	Centimeters	100.0
Meters	Feet	3.281
Meters	Inches	39.37
Meters	Kilometers	0.001
Meters	Millimeters	1,000.0
Meters	Yards	1.094
Meters/minute	Feet/minute	3.281
Meters/minute	Kilometers/hour	0.06
Meters/minute	Knots	0.03238
Meters/minute	Miles/hour	0.03728
Meters/second	Feet/minute	196.8

(Continued)

TABLE B-11   Common Conversion Multipliers (*Continued*)

Convert	Into	Multiply by
Meters/seconds	Kilometers/hour	3.6
Meters/seconds	Miles/hour	2.237
Meter-kilograms	Foot-pounds	7.23
Miles	Centimeters	160926
Miles	Inches	63360
Miles	Feet	5280
Miles	Kilometers	1.609
Miles/hour	Kilometers/hour	1.609
Miles/hour	Meters/minute	26.824
Miles/hour	Meters/second	0.447
Millimeters	Centimeters	0.1
Millimeters	Feet	3.281×10 E-3
Millimeters	Inches	0.03937
Millimeters	Kilometers	10 E-6
Millimeters	Meters	0.001

O

Ounces	Grams	28.349527
Ounces	Pounds	0.0625
Ounces (fluid)	Cubic inches	1.805
Ounces (fluid)	Liters	0.02957

P

Pints (U.S. liquid)	Cubic centimeters	473.26
Pints (U.S. liquid)	Cubic feet	0.0167
Pints (U.S. liquid)	Cubic inches	28.877
Pints (U.S. liquid)	Liters	0.4733
Pounds	Grams	453.5924
Pounds	Kilograms	0.4536
Pounds	Ounces	16.0
Pound-feet	Meter-kilograms	0.1383
Pounds of water	Gallons	0.1198
Pounds/square inch	Atmospheres	0.06804
Pounds/square inch	Feet of water	2.307
Pounds/square inch	Inches of mercury	2.036
Pounds/square inch	Kilograms/square meter	703.1

Q

Quarts (liquid)	Cubic inches	57.75
Quarts (liquid)	Cubic meters	9.464×10 E-4
Quarts (liquid)	Cubic yards	1.238×10 E-3
Quarts (liquid)	Liters	0.9463

T

Temperature (°C) +273	Absolute temperature (°C)	1.0
Temperature (°C) +17.78	Temperature (°F)	1.8
Temperature (°F) +460	Absolute temperature (°F)	1.0
Temperature (°F) −32	Temperature (°C)	5/9

TABLE B-11   Common Conversion Multipliers (*Continued*)

Convert	Into	Multiply By
W		
Watts	Horsepower	1.341×10 E-3
Y		
Yards	Meters	0.9141

Compressed Air Horsepower

The horsepower rating for compressors is normally an indicator of the compressed air output. It has, however, become a common practice to rate so-called "commercial," "contractor," or "home" compressors by their motor horsepower. That is, a compressor that has an output of 12 SCFM at 125 psi is a true 2.8-hp compressor, even though the graphics may advertise "6 hp." This 6-hp rating is the horsepower of the motor, not the compressed air output. In an effort to present the image of a higher capacity compressor for less money, many companies boldly advertise the motor hp rather than the air compression horsepower. This has a tendency to skew the customer's perception into thinking that they are getting a lot of compressor for a little price. This is not true and can be a very expensive way to learn about compressed air.

When evaluating any compressor, it is very important to base your evaluations on SCFM (CFM) and not the motor horsepower.

To further illustrate this, let's compare a true 5-hp industrial compressor with a commercial 6-hp compressor available from the local home improvement center.

1. The 5-hp industrial compressor costs $989.00 and produces 16.9 SCFM at 175 psi. $16.9 \times 0.282 = 4.77$ Actual horsepower of compressed air. $989.00 \div 4.77$ hp = $207.34 per hp.

2. The 6-hp commercial compressor costs $229.99 and produces 4 SCFM at 90 psi. 4.0×0.195 (from page III) = 0.78 Actual horsepower of compressed air. $229.99 \div 0.78$ hp = $294.86 per hp! Nearly $100.00 dollars more per horsepower than the industrial compressor!

As you can see, the industrial compressor really is a 5-hp compressor and represents a much better compressed air value. On the other hand, the 6-hp commercial compressor is really a $^{3}/_{4}$-hp compressor! A far cry from what you were probably expecting.

Additionally, many of these lesser compressors have what is termed "compressor-rated motors." This can also be a little confusing to the customer. "Compressor-rated motor" is a relatively new term, which actually refers to a low-duty cycle motor. These motors are not suitable for continuous operation. They are capable of outputting their power for only short times and require a longer "off" period for cool down. Low-duty cycle motors are considerably less expensive to manufacture and have the added advertising advantage of a high hp rating.

Coupled with the lesser actual compressed air output and the low-duty cycle motor, it is not difficult to imagine that the compressor may be forced to operate well above its maximum rated capabilities. This will have the effect of greatly reducing the life of your compressor. Once again, a costly education in compressed air.

Remember, just because the compressor has a 5-hp motor, it doesn't mean that it'll produce 5 hp of compressed air. Check the SCFM rating.

Horsepower Required to Compress Air

The following chart is intended as a general guide for the average horsepower that is required to compress air under normal operations. Actual horsepower required for any given compression application must be determined with the equipment that is being utilized and in the environment in which the system must operate.

Horsepower required to compress one SCFM. (Horsepower plus 15 percent to compensate for inefficiencies.)

Required

Psi	hp	Psi	hp
10	0.04	90	0.20
20	0.07	100	0.21
30	0.10	125	0.23
40	0.12	150	0.26
50	0.14	175	0.28
75	0.18		

Occupational Safety & Health Administration (OSHA) Laws and Regulations Concerning Compressed Air*

- Part Number: 1926
- Part Title: Safety and Health Regulations for Construction
- Subpart: I
- Subpart Title: Tools—Hand and Power
- Standard Number: 1926.306
- Title: Air receivers.

1926.306(a)

"General requirements."

1926.306(a)(1)

"Application." This section applies to compressed air receivers, and other equipment used in providing and utilizing compressed air for performing operations such as cleaning, drilling, hoisting, and chipping. On the other hand, however, this section does not deal with the special problems created by using compressed air to convey materials nor the problems created when men work in compressed air as in tunnels and caissons. This section is not intended to apply to compressed air

---

*Web address: www.osha.gov.

machinery and equipment used on transportation vehicles such as steam railroad cars, electric railway cars, and automotive equipment.

1926.306(a)(2)

"New and existing equipment."

1926.306(a)(2)(i)

All new air receivers installed after the effective date of these regulations shall be constructed in accordance with the 1968 edition of the A.S.M.E. Boiler and Pressure Vessel Code Section VIII.

1926.306(a)(2)(ii)

All safety valves used shall be constructed, installed and maintained in accordance with the A.S.M.E. Boiler and Pressure Vessel Code, Section VIII Edition 1968.

1926.306(b)

"Installation and equipment requirements."

1926.306(b)(1)

"Installation." Air receivers shall be so installed that all drains, handholes, and manholes therein are easily accessible. Under no circumstances shall an air receiver be buried underground or located in an inaccessible place.

1926.306(b)(2)

"Drains and traps." A drain pipe and valve shall be installed at the lowest point of every air receiver to provide for the removal of accumulated oil and water. Adequate automatic traps may be installed in addition to drain valves. The drain valve on the air receiver shall be opened and the receiver completely drained frequently and at such intervals as to prevent the accumulation of excessive amounts of liquid in the receiver.

1926.306(b)(3)

"Gages and valves."

1926.306(b)(3)(i)

Every air receiver shall be equipped with an indicating pressure gage (so located as to be readily visible) and with one or more spring-loaded safety valves. The total relieving capacity of such safety valves shall be such as to prevent pressure in the receiver from exceeding the maximum allowable working pressure of the receiver by more than 10 percent.

1926.306(b)(3)(ii)

No valve of any type shall be placed between the air receiver and its safety valve or valves.

1926.306(b)(3)(iii)

Safety appliances, such as safety valves, indicating devices and controlling devices, shall be constructed, located, and installed so that they cannot be readily rendered inoperative by any means, including the elements.

1926.306(b)(3)(iv)

All safety valves shall be tested frequently and at regular intervals to determine whether they are in good operating condition.

- Part Number: 1910
- Part Title: Occupational Safety and Health Standards
- Subpart: P
- Subpart Title: Hand and Portable Powered Tools and Other Hand-Held Equipment
- Standard Number: 1910.242
- Title: Hand and portable powered tools and equipment, general.

1910.242(a)

General requirements. Each employer shall be responsible for the safe condition of tools and equipment used by employees, including tools and equipment which may be furnished by employees.

1910.242(b)

Compressed air used for cleaning. Compressed air shall not be used for cleaning purposes except where reduced to less than 30 psi and then only with effective chip guarding and personal protective equipment.

- Part Number: 1915
- Part Title: Occup. Safety and Health Standards for Shipyard
                         Employment
- Subpart: H
- Subpart Title: Tools and Related Equipment
- Standard Number: 1915.131
- Title: General precautions.

The provisions of this section shall apply to ship repairing, shipbuilding, and ship breaking.

1915.131(a)

Hand lines, slings, tackles of adequate strength, or carriers such as tool bags with shoulder straps shall be provided and used to handle tools, materials, and equipment so that employees will have their hands free when using ship's ladders and access ladders. The use of hose or electric cords for this purpose is prohibited.

1915.131(b)

When air tools of the reciprocating type are not in use, the dies and tools shall be removed.

1915.131(e)

Before use, pneumatic tools shall be secured to the extension hose or whip by some positive means to prevent the tool from becoming accidentally disconnected from the whip.

1915.131(g)

Headers, manifolds and widely spaced hose connections on compressed air lines shall bear the word "air" in letters at least 1-inch high, which shall be painted either on the manifolds or separate hose connections, or on signs permanently attached to the manifolds or connections. Grouped air connections may be marked in one location.

1915.131(h)

Before use, compressed air hose shall be examined. Visibly damaged and unsafe hose shall not be used.

Answers to Chapter Questions

Chapter 1

1. C **2.** A **3.** D **4.** C **5.** B
6. A, B, C, D **7.** B **8.** A **9.** B **10.** C

Chapter 2

1. D **2.** B **3.** A, C **4.** B **5.** A, B, C, D
6. D **7.** B **8.** A, B, C, D **9.** A, B **10.** B, D

Chapter 3

1. B **2.** B **3.** C **4.** B **5.** D
6. A **7.** C **8.** D **9.** A **10.** B
11. B **12.** D **13.** B **14.** D **15.** B

Chapter 4

1. A, D **2.** C, D **3.** B **4.** B **5.** A
6. C **7.** C **8.** B **9.** C

Chapter 5

1. A **2.** C **3.** D **4.** A, C **5.** C, D
6. B **7.** A

Chapter 6

1. A, B   **2.** A, D   **3.** A, C   **4.** C   **5.** C
6. B   **7.** D   **8.** C   **9.** C   **10.** B

Chapter 7

1. B   **2.** A   **3.** D   **4.** D   **5.** C, D
6. B   **7.** C   **8.** A   **9.** A, C   **10.** B

Chapter 8

1. A   **2.** C   **3.** B   **4.** C   **5.** A

Chapter 9

1. C   **2.** A   **3.** D   **4.** C   **5.** A
6. A, B, C, D   **7.** C   **8.** D   **9.** B

Chapter 10

1. B   **2.** A   **3.** B   **4.** A, B, D   **5.** C
6. B, D   **7.** C   **8.** C   **9.** A, B, C, D   **10.** B, D

Chapter 11

1. C   **2.** A   **3.** D   **4.** C   **5.** B
6. A   **7.** C   **8.** B   **9.** B

Chapter 12

1. B   **2.** D   **3.** A   **4.** A, D   **5.** C
6. A   **7.** C   **8.** D   **9.** B   **10.** D

Chapter 13

1. C   **2.** B   **3.** D   **4.** A, B   **5.** A
6. B, C   **7.** D   **8.** B   **9.** B   **10.** D
11. B   **12.** D   **13.** B   **14.** D   **15.** A
16. C   **17.** D   **18.** A   **19.** B

Chapter 14

1. B   **2.** D   **3.** A   **4.** B   **5.** A, B, C, D
6. A, C, D

Index

120 VAC control loop, 228
24 hour/7 day controller, 104

Absorption dryer, 157
AC power, 224
After coolers, 104
Air
   audit, 341
   audit outline, 350
   bag jack, 59
   bearings, 52
   brakes, 62
   brush, 38
   chuck, 12
   cooled after coolers, 106
   cylinder, 16
   dryers, 108
   electricity, 277
   end, 87, 101
   hammer, 28
   over hydraulic jack, 56
   motor, 29
   quality monitoring, 310
   ratchet, 34
   ride suspension, 61
   tank, 14
   tool performance, 275
   wrench, 36
Application maintenance, 261
Applications, 9
Approach temperature, 108
Auto service shop, 288
Automatic drain, 120
Automobile lift, 55
Automotive system, 283
Axial compressor, 91

Ball valve, 186
Balloon, 1
Barrel jack, 58

Bellows, 1, 2
Belts, 371
Bends, 4
Bicycle tire pump, 71
Blacksmith shops, 1
Blowgun, 1
Blow-off guns, 10
Blowers, 2
Bolt cutter, 30
Bottle jack, 57
Box stapler, 47
Brooklyn Bridge, 3
Buried pipe, 334
Butterfly valve, 189

Cabinet blaster, 43
Cable drain, 180
Caisson disease, 4
Caissons, pressurized, 4
Canister dryer, 151
Caulk guns, 48
Centrifugal blower, 74
Centrifugal trap, 122
CFC-based refrigerated dryers, 109, 128
Check valve, 170
Chemical plants, 305
Choosing a compression system, 321
Coalescing filters, 125
Coil hose, 174
Commercial pipe fitting, 181
Common conversion multipliers, 391
Compressed air
   applications, 9
   dryers, 117
   history, 1
   horsepower, 395
   supply, 99, 272
Compression fitting, 183
Compression ratio, 95
Compression system, specifying, 321

Compressor, 348
  control switch, 198
  mounts, 202
  room, 260
  selecting, 323
  types, 71
  home built, 355
Conservation of mass and energy, 95
Construction compressor, 115, 312
Continuous run unloader valves, 199
Continuous run, 99
Contractor compressor, 6, 7, 281
Cost savings, 329
Costs, 269
Crow foot, 177
Cycling dryer, 132
Cyclone trap, 122
Cylinder 16
  control, 19, 216
  control station, 291
  force, 23, 24

Daily duplex operation, 250
Decimal equivalents, 389
Deliquescent dryer, 154
Delta/wye starter, 233
Dental compressor, 89
Dental drill, 38
Dental system, 296
Desiccant dryer, 110, 150
Diaphragm compressor, 78
Diaphragm cylinder, 17, 20
Die grinder, 10, 33
Diesel-powered compressor, 115
DIN rails, 252
Discharge temperature, 272
Disk sander, 34
Distribution
  layouts, 331
  maintenance, 259
  manifold, 178
  map, 347
Diving compressor, 3
Diving, 2
Double-acting compressor, 80
Double-acting cylinder, 18
Double-ended cylinder, 18
Double pole control, 226
Drain valves, 178
Drill tap sizes, 388
Drive belt, 371
Drive belts, 274
Drop trap, 118

Dry cleaning operations, 291
Dryers, 348
  absorption, 157
  applications greater than 120
    SCFM, 144
  CFC-based, 109, 128
  deliquescent, 154
  desiccant, 110, 150
  duplex installation, 142
  dry ice, 365
  Elliott cycle, 113, 135
  home built, 365
  membrane, 154
  performance, types and
    applications, 157
  refrigerated, 109, 128
Dual diaphragm pump, 50
Dunlop, John, 4
Duplex compressor, 101
Duplex controller, 103, 246
Duplex dryer installations, 142
Duty cycle, 322

Electric motor, 273
Electrical controls, 223
Electricity verses air, 277
Electronic drain, 179
Elliott cycle dryer, 113, 135
Elliott refrigeration cycle, 136
Emergency stop, 240
Employee training, 349
Enclosures, 223
Energy savings, 269, 275
Estimating SCFM, 322
Expansion joints, 172, 208
External seat drain valve, 179

Field compressors, 113
Field distribution system, 314
Field dryer, 314
Field systems, 312
Filters, 164, 345
Fire extinguisher, 15
Fireplace bellows, 1, 2
Float drain, 179
Floor jack, 59
Foot-operated bellows, 1
Frequency, 224
FRL, 167

Gasoline-powered compressor, 115
Gate valve, 186
Gauge, 196

Globe valve, 189
Grease gun, 48
Guide to common leaks, 271

Hand drill, 34
Hand-operated bellows, 1, 2
Hand piece, 38
Heat, 272
Heating, ventilation, and air
    conditioning, 293
Heavy duty air drill, 35
Hertz, 224
High-pressure compressor, 315
High-purity work station, 309
High purity, 305
High-temperature CFC-based air
    dryers, 130
History, 1
Home built compressor, 355
Home built dryer, 365
Home compressor, 6
Home shop, 298
Horsepower required to compress air, 396
Hose, 173, 345
    clamps, 175
    connectors, 173
    connectors, 176
    fittings, 177
    maintenance, 261
    terminations, 270
HVAC, 293
HVAC, 6
HVLP paint gun, 39
Hydraulic pump, 27

Ice bath dryer, 146, 365
Impact wrench, 36
Industrial quick disconnect, 176
Injector compressor, 92
Instrument air systems, 144
Instrument air, 110, 302
Intake filters, 195
Intermittent run, 99

Jack hammer, 30

Large manufacturing facilities, 298
Leaks, 270, 342, 343
Lift control, 286
Line trap, 120
Liquid ring compressors, 77
Lock out/tag out program, 243, 257
Lock out/tag out tools, 246

Loop distribution system,
    303, 332
Lubricators, 166, 346
Lungs, 1

Maintenance, 255
    calendar, 262
    items, 258
    program, 255, 262
    time table, 263
Manufacturing plants, 288
Master regulator, 274, 324, 347
Membrane dryer, 154
Micro grid coalescing filter, 125
Microprocessor, 240
Mining, 3
Misuse, 272, 330
Motor starter, 227
Mufflers, 195
Multiple pipe sizes, 372

Nail gun, 45
National Electrical Manufacturers
    Association, 223
Needle scaler, 28
Needle valve, 193
NEMA motor frame sizes, 383
NEMA, 223
Nitrox, 316

Oil/water separators, 158
On, off and automatic mode, 231
OSHA regulations, 397
OSHA, 257

Packaged compressor, 99
Paint booth, 285
Paint gun, 39
Paint sprayers, 38
Parallel duplex operation, 249
Paris, compressed air system, 3
Peak load, 102
Percussion tools, 27
PID, 218
Pilot valve, 192
Pipe
    areas, 380
    buried, 334
    diameters, 380
    fitting dimensions, 378
    fitting, 181, 373
    flow, 326
    hanger, 184

Pipe (*Cont.*):
  size, 326
  sizes, schedules, and wall thickness, 381
  tap drill sizes, 387
  thread engagement, 377
  threading, 338
  vise, 337
  weight per foot, 382
Piston compressor, 79
Plant floor work station, 290
Plate separators, 124
PLC, 238
Plug valve, 180
Plumbing tools, 336
Plumbing, 333
Pneumatic controls, 211
Pneumatic suspension, 60
Pneumatic symbols
  in house, 217
  standard, 212
Pneumatic tire, 4
Pop, Viktor, 3
Pop riveter, 49
Porous media coalescing filter, 127
Portable air tank, 14
Power disconnects, 243
Pressure
  amplifier, 25
  conversion scale, 375
  gauge, 196
  regulators, 165
  release valve, 168
  relief valve, 169
  switch, 198
Pressure/temperature ratings, 373
Pressurized fire extinguisher, 15
Primary receiver, 347
Production paint gun, 40
Production time, 278
Programmable logic controllers, 238
PVC pipe, 334

Quick disconnect, 176

Receiver size, 325
Receiver, 347
Recommended pipe sizes, 328
Redundant compressor, 103
Redundant system, 300
Regenerative blower, 75
Regenerative dryer, 110, 152
Regulator, 165, 345
Reheaters, 130

Remote receivers, 327, 346
Replacement parts inventory, 264
Retractable hose, 174
Rivet cutter, 30
Roebling, Washington, 4
Roots blower, 77
Rotary lobe blower, 77
Rotary vane air motor, 29
Rotary vane compressor, 88
Rotating tools, 33
Routine maintenance, 349

Safety blow off gun, 11
Safety valve, 168
Salt water prepared dryers, 145
Sand blasters, 42
SCBA, 68, 315
SCFM requirements, 343
SCFM, 322
Schrader valve, 195
SCR, 240
Screw compressor, 85
Screw compressor, 99
SCUBA, 66, 315
Self-contained breathing apparatus
  (*see* SCBA)
Self-contained underwater breathing
  apparatus (*see* SCUBA)
Sensor control loop, 229
Sensors, 232
Service outlet, 243
Service tank, 14
Service truck, 310
Sheave location and alignment, 376
Sign-off sheets, 262
Single-pole control, 225
Siphon guns, 40
Sizing a receiver, 325
Skid mounting, 203
Soft starter, 235
Solenoid valve, 191
Sound damping enclosure, 201
Speed control valve, 194
Spine distribution system, 292, 331
Spray mister, 53
Spring return cylinder, 18
Squirrel cage blower, 73
Stand-alone after cooler, 108
Standard dryer installations, 141
Standard voltages, 224
Stapler, 46
Steam, 2
Straight NPT connector, 176

Strainer, 181, 182
Supply maintenance, 256
Support components, 153
System examples, 281
System pressure, 274

Tap drill sizes, 388
Tap drill sizes, pipe, 387
Taper pipe thread engagement, 377
Temperature conversion formula, 374
Temperature conversion scale, 374
Tension, 371
Thompson, Robert, 4
Time savings, 278
Tire gauge, 13
Tire inflators, 12
Tire pump, 4, 5, 71
Tire valve, 195
Traps, 118
Tumble blaster, 43
Turbine motors, 37
Turbo compressor, 90
Twin tower dryer, 152
Two-stage compressor, 4, 6, 83

Universal hose coupling, 177
Unloader valve, 199

Unloader, 101
Utility trunk, 306

Vacuum generator, 51
Valve stems, 270
Variable frequency drive, 237
V-belt, 200, 274
    tension, 371
Venting, 255
Vibration isolator, 172
Vibration mounts, 202
Vibrators, 54
Von Guericke, Otto, 2

Water buildup, 117
Water contamination, 117, 274
Water-cooled after cooler, 107, 128
Water damage, 274
Water-powered bellows, 1
Weekly duplex operation, 250
Wire guide, 252
Wire types & gauges, 251
Work station, 285
World War II, 5

Y-Strainer, 181